UNITED NATIONS PUBLICATION
Sales N° 90.III.D.1
ISBN 92 807 1253 5

03200P

not subject to official editing by the UNEP/... conferences and Governing Council Service

First edition 1990

The designations employed and the presentation of the material in this publication do not imply the expression of any opinion whatsoever on the part of the United Nations Environment Programme concerning the legal status of any country, territory, city or area or of its authorities, or concerning the delimitation of its frontiers or boundaries. Moreover, the views expressed do not necessarily represent the decision or the stated policy of the United Nations Environment Programme, nor does citing of trade names or commercial processes constitute endorsement.

UNITED NATIONS PUBLICATION

Sales N° 90.III.D.1

ISBN 92 807 1253 5

FOREWORD

"Environmental Auditing" is a management tool used by industry to evaluate its environmental performance. However, while the use of this tool is growing, too few companies still are using it. This is why, in order to contribute to the exchange of information on the subject, the United Nations Environment Programme/Industry and Environment Office (UNEP/IEO) convened a ~~p.198.~~ senior-level working group of experts from companies where environmental auditing is currently in use in Paris on 10 and 11 January 1989.

The present document summarizes the discussions and gives the full text of the various presentations, copies of public guidelines issued in the United States and a short list of references. It is clear from the presentations and the discussions that the concept of "Environmental Auditing" covers various methodologies and various approaches. No one system can work well for everybody; instead each company has to define its own system, depending upon its size, its activity, its own "culture". But there is now general agreement that environmental auditing, when performed in a proper manner, is a means for a company to assess its environmental performance and to improve the effectiveness of its environmental policy.

UNEP/IEO hopes that this document, which provides concrete examples and case studies of current environmental auditing practices, with different degrees of formality and complexity, will encourage industry managers to use this new tool.

ACKNOWLEDGEMENTS

UNEP/IEO wishes to acknowledge the most valuable participation in the workshop of the following experts:

- Mr. Richard Almgren, Federation of Swedish Industries
- Mr. Nigel Blackburn, International Chamber of Commerce
- Mr. Eric B. Cowell, BP International Ltd.
- Dr. G. Eigenmann, Ciba-Geigy AG
- Dr. I.J. Graham-Bryce, Shell Internationale Petroleum
- Dr. Egon Keller, Ecosystem
- Mr. Michael L. Kinworthy, Unocal Corporation
- Mr. Jonathan Plaut, Allied Signal Inc.
- Mr. Jacques Salamitou, Rhône-Poulenc
- Mr. Per A. Syrrist, Norsk Hydro A.S.
- Mr. Hennie Veldhuizen, Noranda Inc.
- Ms. Linda A. Woolley, ITT Corp.

UNEP/IEO was represented by:
- Mrs J. Aloisi de Larderel, Director
- Mr. T. de la Perrière, who, on a voluntary basis, acted as rapporteur of the meeting.

UNEP/IEO further wishes to acknowledge the support of the French Ministry of the Environment (Service de la Recherche, des Etudes et du Traitement de l'Information sur l'Environnement) in this activity.

TABLE OF CONTENTS

CHAPTER 1 - SUMMARY OF DISCUSSION

At the invitation of the United Nations Environment Programme/Industry and Environment Office (UNEP/IEO), twelve senior-level industry experts participated in a workshop on "environmental auditing".

Each of the participating experts presented the current method of planning, conducting and using an environmental audit in their own company. In addition, the Chairman of the International Chamber of Commerce (ICC) Group on Environmental Auditing presented the ICC position paper on the subject. The list of participants and the text of each presentation are given hereafter. Each of the presentations was followed by a discussion, and a general discussion took place during the last afternoon.

The following summarizes the main points of the discussions.

I - DEFINITION AND BENEFITS FROM ENVIRONMENTAL AUDITING

There was general agreement on the definition of environmental auditing adopted by the ICC:

"A management tool comprising a systematic, documented, periodic and objective evaluation of how well environmental organization, management and equipment are performing with the aim of helping to safeguard the environment by:

(i) Facilitating management control of environmental practices;

(ii) Assessing compliance with company policies, which would include meeting regulatory requirements".

However, even if the expression "environmental audit" is now acknowledged by most companies, some prefer to use other terms such as "environmental surveillance" (Allied-Signal), "environmental review" (Unocal), "environmental quality control" (Rhône-Poulenc), or "environmental assessment".

Participants stressed the benefits which can be obtained from environmental audits.

The primary benefit of an audit is that is ensures cost-effective compliance not only with laws, regulations and standards, but also with company policies. Many companies have now set up internal environmental standards and guidelines that they are applying world-wide. These may be more stringent that local regulations and laws. In addition environmental auditing can have many associated benefits, such as increasing management and employee awareness of environmental issues, thus leading to better overall environmental management.

In summary, environmental audits should provide answers to the following questions raised by company managers:

- what are we doing? In particular, are we in compliance with government regulations, guidelines, codes of practice, permit consitions?
- can we do it bettter? In particular, are there non-regulated areas where operations can be improved to minimize the impact on the environment?
- can we do it more cheaply?
- what more should we do?

All participants stressed that environmental auditing is a management tool which is valuable only if it is moulded in an overall management system. It cannot stand alone. It is a monitoring tool which aids decision making and management control. Of course, government inspections to ensure compliance with the law have to be performed.

The participants felt that environmental audits should remain an internal process and that results should not be disclosed outside the company, for the following reasons:

- in order to maintain a relationship of trust between auditors and auditees and avoid any hiding of inadequacies,
- because environmental auditing goes beyond compliance appraisal, requires open discussion of confidential business and operational information, and can even involve strategic consideration of ways to improve operations,
- because auditing processes are evolving and being improved continuously and would run the risk of becoming "fossilized" if strictly regulated.

One participant, however, suggested the possibility of involving authorities in part of the auditing process. It was generally agreed that the audits should be made available in the case of an accident and litigation. Experts from the USA and Canada recalled the policies of their Governments.

Participants also recognized that public information is absolutely necessary to enable dialogue between the plant manager and the people living in the vicinity. A plant with a fence around it may seem threatening; but much of the mystery disappears if the community is informed of its potential hazards, effluents, wastes and other potential effects on the environment. Environmental auditing should remain separated from provision of public information, but it will certainly provide elements which can be drawn in keeping the public fully informed.

II - PLANNING FOR ENVIRONMENTAL AUDITING

II.1 - General organization

In most companies, environmental audits are the responsibility of a specialized unit, within the environment department at company headquarters. This independent, centralized unit is generally also in charge of health and safety audits. The two types of audits are normally handled separately, however, as they require different skills.

It should also be noted that, in addition to the audits prepared under the responsibility of the specialized unit, in some companies, each plan within a company generally has to prepare annually an "internal environmental audit" to be sent to headquarters.

All participants stressed that environmental auditing needs the strong endorsement and active support of top management, and that the procedure should be clearly communicated together with the appropriate incentives.

Also, all participants stressed that full measures should be taken to gain the confidence of the auditees: the audit is performed to help improve future environmental performance, to identify ways to progress, and not to punish the managers. For most companies, the audits should not serve as a basis for comparison of the plant's managers. For others, comparison between the performance of the plants could be an incentive to improve the environmental performance and policies.

There should not be any element of surprise. On the contrary, plant management and staff should be informed well in advance of the date, objectives, and suggested protocol of the audit. At Ciba Geigy, for example,a letter is sent well ahead of time. This contributes to creating a climate of trust.

II.2 - Selection of facilities to be audited

Most companies audit all of their facilities regularly, in all countries where they are located, whatever the size of the facility (in some cases the facilities might be operated by only one to two employees).

However the audits are more or less frequent according to the degree of hazard linked to the facility.

Unocal, for example, audits each refinery every 18 months medium-risk facilities every three years, and low-risk facilities every five to six years. In each risk category, facilities to be audited each year are generally selected randomly. But where a need for substantial improvements is identified, a particular facility may be audited more frequently than the average.

BP audits not only facilities, but also activities in general (for example, shipping) and how environmental issues (for example, hazardous waste disposal) are being dealt with. Similarly, Unocal is undertaking an assessment programme of its vulnerability and capability of responding to oil spills and hazardous materials realeases. In addition, not only Unocal audits its own facilities, but also audits hazardous waste disposal sites that are used by company operations

II.3 - Composition of the auditing team

Participants consider the selection of the auditing team members as critical for the success of an audit. The team is generally composed of two to eight people.

Auditors may be full-time professionals, subject specialists or representatives from the business unit being audited. Or they may come from similar units or be qualified external consultants. They are usually selected according to the nature of the audited unit, and sometimes only after the key issues have been identified. In any case the team should include both environmental experts, and people familiar with the type of operation being audited.

The role of the team leader is essential. Generally the team leader is a professional member from the central environmental or health and safetydepartment, but this is not always the case. At Norsk Hydro or Noranda,

for example, the team leader can be the production manager of one of the operating facilities other than the one being audited.

External consultants can be part of the team. Allied-Signal and ITT, for example, Noranda and Rhône-Poulenc exceptionally or never do so. Among the benefits of outside consultants' participation were mentioned their external experience, their knowledge of local laws, and in some cases easier acceptance by the auditees. They provide a "pair of fresh eyes". Some participants, on the other hand, indicated that communication barriers can be created if staff are not accustomed to interacting with external consultants.

The general consensus, in conclusion, is that consultants can be useful but are not indispensable for large companies. Small or medium-size companies, which do not have internal environmental specialists, could use them more often.

The participation of a representative from the plant being audited is considered useful by some companies (Shell, Ciba-Geigy) because of his/her knowledge of the plant, the key people to meet, etc. and the fact that, "protected" by the team, this person can more easily be critical. For the sake of objectivity, other companies prefer not to include a plant representative on the team, but to keep in permanent contact with one (usually the plant's environmental manager).

Some companies (Noranda, Shell, Norsk Hydro, Unocal) include in the auditing team representatives from other company plants with similar activities. This interchange is a powerful tool for the exchange of experience and training, and contributes to ensuring the comparability of audits and consistency of approach.

III - CONDUCTING THE AUDIT

Each company has developed its own methodology and set of procedures for conducting an audit. Site visits, questionnaires, interviews, and review of documentation are used. Companies have also prepared checklists of points to be inspected. Well-defined and systematic procedures, known and understood by all concerned, are necessary for the quality, consistency and comparability of the results as well as for the smoothness of the process.

The ICC presentation on environmental auditing describes the basic steps in an environmental audit.

Books have been published and training courses are now organized for environmental auditors. A few publications are listed at the end of this document.

All participants considered that the following elements need to be audited:
- policy and principles
- systems
- procedures
- practice
- performance

Rhône-Poulenc described its work to quantify overall performance with an environmental index, and the use of audits which permit tracing the improvement of each plant index over the years.

IV - USING THE AUDIT: REPORTS AND FOLLOW-UP

All participants underlined that the audit should be properly and quickly documented in written reports, concentrating on factual and objective observations, after discussion with facility personnel and management.

Usually a draft report is prepared and sent to the site management. The final report is established only after the site management has had the opportunity to comment. It is then distributed to management, in some cases to all employees who are involved in the audit on the sites.

For many companies (Allied-Signal, Shell, ITT, Noranda) the report should contain only an analysis of the situation, underlying strengths and weaknesses

and proposing deadlines for action. It should not contain suggestions or recommendations on how to solve the final problems. It is the task of the plant management to define the appropriate measures and means to be used. However, at the request of plant managers, headquarters can provide advice or appropriate information. In other companies (Rhône-Poulenc), the measures to be taken are jointly defined by the audit team and the plant management, and are part of the report.

In any case, the value of the whole process obviously depends on the effective and timely action to address problems identified. For that reason Allied-Signal, for example, conducts a number of one-day follow-up reviews one year after the main audit. Unocal performs quarterly updates to see if the schedule is being followed.

In some companies, the activities of the audit unit are reported to the Board, as is the case for example for Allied-Signal (every four months) and Noranda and Unocal(every year).

V - ASSURING THE QUALITY OF ENVIRONMENTAL AUDITING

Assuring the permanent high quality of environmental auditing is regarded as essential.

The most important element in achieving this objective is considered to be the choice of team members. All participants emphasized strongly on the fact that auditing is a very demanding and stressful job, requiring long working hours on-site. Auditors must have high professional competence, which takes a long time to acquire. They must have psychological skills and the diplomacy necessary to adapt to the different types of auditees. They should be open to dialogue and at the same time keep their objectivity.

To ensure the competence and efficiency of auditors, all participants agreed that a lot of attention must be given to their training. At Noranda and ITT for example, extensive training is provided by a specialized consulting firm, including a trial audit, and a four-day yearly refresher course is organized.

VI - CONCLUSION

All participants agreed that environmental audits were beneficial to both the environment and the companies themselves.

Most of the case studies presented came from large companies. The question of the applicability of environmental audits in small and medium-size companies, and in developing countries, was raised.

Experts considered that environmental auditing is a universal concept, applicable world-wide. However, methodological support could be provided to small and medium-size companies through industry/trade associations. The Swedish Federation of Industries, for example, is considering providing such support.

In developing countries, training activities and support could be provided through UNEP/IEO, the ICC and the International Environmental Bureau (IEB), a division of the ICC.

Finally all present recognized that environmental auditing does not alter the necessity for companies to respond to government reporting requirements and compliance inspections. There was also clear evidence that, although environmental auditing should remain basically an internal management tool, companies have to respond to public information requests.

The need for more work on these themes was underlined.

CHAPTER 2 - ALLIED-SIGNAL INC. HEALTH, SAFETY AND ENVIRONMENTAL SURVEILLANCE PROGRAM

Jonathan PLAUT
Allied-Signal Inc.

The impetus for Allied-Signal's Health, Safety and Environmental Surveillance program came in 1978 when a special committee of the Board of Directors approved a recommendation by its outside consultant to develop an environmental auditing program. This recommendation followed a corporate-wide assessment of the health, safety and environmental status of its operations. In response to that recommendation, the corporate health, safety and environmental affairs staff, the Corporate Audit Department, the Law Department, and the consultant, with the cooperation of business and operating managers, devised a program called the Health, Safety and Environmental Surveillance Program.

Allied-Signal's audit program is housed as a separate activity within the Corporate Health, Safety and Environmental Sciences Department. There are three full-time environmental auditors who comprise the Health, Safety and Environmental Surveillance staff. Each audit team includes at least one member of the Surveillance staff, an outside consultant, and may include a corporate or operating company environmental

---

Jon Plaut is Director of Environmental Compliance of Allied-Signal, with responsibility for coordination and management of the Corporation's international health, safety and environmental activities. He has degrees in law and engineering, and a Masters Degree in international law. He has published articles in many periodicals and is a frequent university and environment seminar lecturer

professional. The Director of the Surveillance Program and the consultant routinely report on the status of the program to the Allied-Signal Board of Directors.

BACKGROUND

The corporation, with about 240 manufacturing facilities and 110.00 employees world-wide, is organized into three sectors: Automotive, Aerospace, and Engineered Materials.

The Corporate Health, Safety and Environmental Sciences Department is headed by a Vice President who reports to the Senior Vice President, Operations Services. The department has 14 full-time staff organized into the following major disciplines:

- pollution control — responsible for air pollution control, water pollution control, solid and hazardous waste disposal, spill prevention and water supply;
- product safety — responsible for product safety and product integrity programs;
- occupional health — responsible for industrial hygiene;
- loss prevention — responsible for prevention of worker-related and property losses; and
- environmental surveillance — responsible for the environmental auditing function.

Each of the above groups is headed by a director. The responsibilities of this corporate department are overall coordination of and guidance to the sectors and operating companies, program monitoring, and regulatory affairs interaction.

Each sector has environmental, health and safety staff (structured similarly to that of the corporate staff, but without a surveillance function) with day-to-day responsibilities for health, safety and environmental compliance in their sector. Each group is headed by a director who reports at a high level within the sector organization. A dotted-line reporting relationship exists with corporate counterparts. Staff assigned to specific facilities report to facility management but have a dotted-line reporting relationship to sector health, safety and environmental counterparts.

PROGRAM PURPOSE

The objective of the Allied-Signal Health, Safety and Environmental Surveillance Program is to provide independent verification that:

- the Corporation's operations are in compliance with the law and with corporate policies and procedures; and
- systems are in place to insure continued compliance.

Allied-Signal Inc.has a formal corporate health, safety and environmental policy issued by Allied-Signal's Chief Executive and Chief Operating Officers. The policy states that the corporation will establish and maintain programs to assure that applicable laws and regulations are known and obeyed, and will adopt its own standards where laws and regulations may not be adequately protective or where laws do not exist. There are also written corporate guidelines for each discipline (e.g., pollution control, safety and health).

Sectors, companies and individual facilities have developed their own written operating procedures to supplement the corporate guidelines. These procedures address environmental concerns specific to the individual businesses.

ORGANIZATION AND STAFFING

The term "surveillance" is used by Allied-Signal in preference to "audit" to avoid possible confusion with financial audits. Financial auditing has had rules, regulations, and generally accepted standards of practice in place for many years; environmental surveillance is an evolving discipline and, as such, rules, systems and standards do not yet exist.

A number of criteria were established when Allied-Signal Inc.considered an approach to organizing and staffing the Surveillance Program including:

- independence of the audit teams from those responsible for managing corporate and sector environmental programs — yet organizationally located where communication and resolution of problems and conflicts would be most efficient;
- minimizing the full-time staffing commitment to the Surveillance Program — yet having a readily available supply of competent, objective team members, continuity in the conduct of reviews, and long-term accountability for the program.

With those objectives, several options were considered, such as using external auditors; establishing an independent internal group housed within the Corporate Audit Department or the Corporate Health, Safety and Environmental Sciences Department; or using task forces made up of persons drawn from throughout the corporation. Each option was viewed by Allied as having advantages and disadvantages. For example:

- an external auditor would not require the addition of any full-time employees, and would have a high degree of independence — yet would involve relatively higher total costs. Additionally, there could be substantial barriers to coordination and communication;
- an independent group within the Corporate Audit Department would have a high degree of independence from health, safety and environmental program management — but would not have the needed un-

derstanding of subject matter.

- a separate, independent group of full-time auditors within the Corporate Health, Safety and Environmental Sciences Department would afford a good opportunity for communication with health, safety and environmental management — but their independence might be questioned;
- a task force would have broad participation and high flexibility in terms of level of effort and relatively low cost — but carry a potential for loss of continuity and disruption of regular functions.

In order to achieve the best mix of the qualities in the options above, a composite approach was adopted. The Health, Safety and Environmental Surveillance Program was established within the Corporate Health, Safety and Environmental Sciences Department and is staffed by three full-time professionals. To ensure continuity and accountability, the team leader for each audit is one of those three professionals. The remainder of the audit team (which varies from two to four people depending on the review scope and size of facility) is comprised of corporate and sector health, safety and environmental professionals familiar with the review subject but not directly involved in the programs being reviewed; and an outside consultant (who provides the advantages of an external auditor).

An audit typically takes three to four days. The current yearly budget for the program is $750,000.

AUDIT SCOPE AND FOCUS

All of Allied-Signal's facilities worldwide are within the scope of the Surveillance Program — with operations assessed as having lower health, safety end environmental risk receiving less attention than those assessed as high-risk operations. The functionnal scope of the program includes:

- air pollution control;
- water pollution control and spill prevention;
- solid and hazardous waste disposal;
- occupational health;
- medical programs;
- safety and loss prevention;
- product safety.

The compliance scope includes all federal, state and local environmental regulations; corporate policies and procedures, and "good health, safety and environmental practices".

With approximately 240 locations and seven functional scopes, a comprehensive review of all subjects at all locations would be a formidable task. Thus, most audits are limited to only one of the functional areas listed above (e.g., air pollution control only) in order to maximize the amount of in-depth review in the time available. Occupational Health and Medical Program audits are conducted simultaneously. Thus, while the program covers a variety of topics, the scope of a specific audit is narrow.

AUDIT TIMING AND FREQUENCY

Approximately 50 audits are conducted annually. The Surveillance Program reviews only a relatively small sample of the Corporation's facilities each year. Review locations are chosen to represent a cross section of Allied-Signal's business interests and health, safety and environmental concerns where the potential risk is high. Corporate environmental staff, the Director of Surveillance, and the external auditor annually develop a sample of facilities to be audited that includes facilities from each of the operating companies and major business areas. Facilities are selected randomly through a process that reflects their assessed environmental risk. Audits are apportioned evenly among the various functional areas.

At the beginning of each year, the audit schedule for the year is sent to corporate and sector health, safety and environmental staff. One month prior to the review, the team leader sends a letter to the facility manager with copies to the appropriate sector and corporate environmental staff.

AUDIT METHODOLOGY

The Health, Safety and Environmental Surveillance Program's comprehensive audit employs a number of techniques such as formal internal control questionnaires, formal audit protocols (or guides), informal interviews with facility personnel, physical observations, documentation review, and verification.

A written audit protocol, which has been prepared for each functional area of the review, reflects the objectives of the audit, facility characteristics, and time constraints. The protocol methodically guides the auditor to an understanding of the management system through the conduct of tests which either confirm that the system is working or determine any specific deficiencies. Auditors carefully document the accomplishment and results of each audit protocol step in their audit working papers.

The basic phases in Allied-Signal's audit process (illustrated in Exhibit 1) include:

Phase I: Preparation

Among the pre-audit activities conducted by the audit team leader (full-time surveillance professional) are the confirmation of review dates and organization of the audit team based on the functional scope of the review. One month in advance of the audit, the team leader notifies the facility manager in writing of the specific dates and review scope. Corporate files are screened to obtain and review information on the facility and its processes (e.g. process flow diagrams, plant layout diagrams, policies and procedures, operating manuals, permits, etc.). Regulations applicable to the facility are also obtained.

Phase II: On-site review

The on-site review commences with a meeting of the audit team, the facility manager, and appropriate facility personnel. During this meeting, the team leader discusses the objectives of the Surveillance Program and the review scope. This is followed by the facility personnel presenting an overview of the facility's operations — products, processes, facility organization, etc. The review team then tours the facility, with a member of the facility environmental or production staff, to gain a general understanding of facility characteristics.

Following the tour, the review team and appropriate facility environmental staff meet to complete the Internal Controls Questionnaire. This questionnaire, administered by the audit team leader, aids the auditors in developing an initial understanding of facility operations, processes, personnel responsibilities and environmental management controls.

Working from the audit protocol, with major sections divided among the audit team members, each auditor gathers system information and performs relevant tests. In the course of the review, the auditor must use sampling techniques and exercise professional judgment in selecting the type and size of samples to be used to verify that the key controls in the system under review are in place and working. No testing may be done until the system is well understood and a carefully reasoned plan of testing is worked out. Such understanding may come from interviews with facility staff, review of facility operating procedures and systems, etc.

Testing of the systems in place can take a variety of forms. For example, verification testing for water pollution control can include:

- visual observation downstream from an outfall;
- comparison of strip charts and discharge monitoring reports;
- review of programs to assure the reliability of treatment or monitoring equipment; and
- determination that composite samplers, effluent flow measuring devices, and in-place monitoring devices are properly maintained and calibrated.

Each auditor carefully documents all testing plans and the results of each test. Each auditor shares his or her observations and information on deficiencies found throughout the audit. Also, time is set aside at the end of each day to exchange information on the system and share any concerns about its effectiveness. The audit team is instructed to continuously feed back any impressions being formed about the system's compliance with established criteria. This continuous feedback is intended to: eliminate misconceptions and false trails for the team member who may have misunderstood what he or she was originally told; encourage the team members to organize their thoughts; and give facility personnel an opportunity to participate in the audit process.

Significant findings are listed by each audit team member and are organized by the team leader on a summary sheet for discussion with facility management. The on-site audit concludes with a close-out meeting between the audit team and facility management. Each receives a copy of the audit findings summary form and each finding is discussed.

Phase III : Reporting and record preservation

Report format

The purpose of the written report is to provide information to top management (sector presidents) on the more significant findings of the audit. The overall thrust of the written report is an opinion as to whether or not the facility is in substantive compliance followed by a list of exceptions noted. The report is based on findings listed on the audit findings summary form. Findings related to regulatory standards are qualified with a statement that they have not received a detailed legal review.

A standardized format for the written report has been established which consists of four parts. Section I is the who, what, where, why information. The next two sections include all significant instances of non-compliance with:

- Regulatory Standards (Federal, state/provincial, and local); and
- Allied-Signal Inc.'s Policies and Procedures (corporate, sector or facility).

The final section includes any significant deficiencies in the facility control systems which would make continued compliance with the law or company policy questionable (such as record retention, documentation, clear assignment of environmental responsibilities, etc.).

The length of the report depends on the number of findings; typically it is four to five pages long.

Report distribution

The written audit report is issued in draft form by the team leader with copies to the involved line and staff personnel at both the operating company and corporate levels, the facility manager and the audit team. Comments on this draft report are requested within two weeks of its issuance. When comments necessitate significant revision of the first draft, a second draft on the report may be prepared and circulated for review.

The final written report is issued to the sector president approximately one month after the review, with copies to the Law Department, Vice President, Health, Safety and Environmental Sciences, corporate environmental functional specialists, business area management, facility manager, and the review team. The final report is accompanied with a request that the operating company respond in writing with an action plan for correcting the deficiencies noted.

Records retention

The Surveillance Department has established a formal records retention policy which was developed to help keep the records volume at a manageable level and to ensure that all records relating to surveillance reviews are retained for a period of time consistent with their utility in the program and with applicable federal regulations. Thus, audit working papers are retained for three years. Audit reports and action plans are retained for ten years (50 years where subject to RCRA or CERCLA records' retention requirements).

Other reporting

In addition to the formal written report on each individual review to the sector president, the Board Corporate Responsibility Committee receives regular reports on the Surveillance Program activities twice a year. The Director of Surveillance attends the meetings of the Board Committee and supplements the written report with an oral report and responds to any questions the Board members may have. Also attending the Board meeting is a representative from the Surveillance Department's outside consulting firm. The purpose of reporting to the Board is to confirm that the Surveillance Program is functioning and to provide assurance that no material deficiencies have been noted.

Phase IV: Company Action

The job of the audit team ends with the submission and management's understanding of the surveillance report. The review process, however, continues until those responsible for correcting any deficiencies noted have prepared an action plan for correcting the deficiencies.

The action plan is developed by facility personnel and sent to the business area manager and the sector health, safety and environmental director. The latter provides a copy of the action plan to the Surveillance Director and other managers of concern. The Surveillance Director receives the action plan and confirms that the final report has been understood and that the response is consistent with the findings of the report.

Action plans are typically received within two to three months of the issuance of the final written report. The plan reports on corrective actions already taken, as well as those that are planned. Operating management then assumes responsibility for follow-up and monitoring of the corrective actions. The Surveillance group performs follow-up reviews to confirm the completion of approximately 20 per cent of these action plans. These follow-up reviews have indicated that most of the findings are corrected in an expeditious manner.

The Corporation's environmental assurance system includes formal procedures for follow-up and corrective acion on all environmental, health and safety deficiencies. A recognition of this commitment is evidenced by the Environmental Assurance Letter which is prepared annually by the sector presidents and submitted to Allied-Signal's Chief Executive and Chief Operating Officers. The letter indicates the state of compliance with Allied-Signal's Health, Safety and Environmental Policy. The objective of the letter is to assure that: (1) appropriate health, safety and environmental systems are in place and functioning; (2) these systems recognize substantial (actual or potential) deficiencies that may exist; (3) such deficiencies are reported up to the necessary level of corporate and sector managements; and (4) appropriate action plans are developed and timely corrective actions taken. The Assurance Letters are reviewed annually at a meeting of the Corporate Responsibility Committee of the Board of Directors.

ASSURING QUALITY

There are a number of ways in which Allied-Signal Inc.ensures audit quality. First, a member of the Surveillance staff participates on each audit as team leader. The team leader's role includes making sure there is good communication among the team members. This is generally done by setting aside a block of time at the end of each day of the audit for the auditors to exchange information and share concerns.

Second, Allied's audit protocols provide a structured framework which guides the auditors through a series of steps designed to create an understanding of the system under review, conduct appropriate tests to confirm that the system is working, and determine specific deficiencies.

A third quality control measure relates to the audit working papers. The credibility of the audit depends on how well each auditor documents what he/she has done and the conclusions reached. Each team member must prepare working papers which document the information gathered in completing the protocol. At the end of each audit, the team leader reviews, initials and dates each page of the working papers. The working papers serve as support for the audit report and a way of evaluating the audit and the performance of each team member.

Finally, Allied-Signal's outside consultant provides an additional quality control check. A representative of the consulting firm participates in each review. All audit reports are reviewed by the consultant to ensure accurate and consistent audit reporting.

PROGRAM BENEFITS AND DEVELOPMENTS

Allied-Signal has noted a number of benefits throughout the Corporation resulting from the Surveillance Program. Among them are the following:

- for top management and the Board of Directors, the program provides independent verification that operations are in compliance with applicable requirements of environmental law and the Corporation's environmental policy;

- for environmental management, the program serves as another source of information on the status of operations, and information on both individual deficiencies and patterns of deficiencies that may occur.

- for line management, there is added incentive for much closer self-evaluation to confirm that operations are in compliance. The program has stimulated line management to become more familiar with the detailed implications of environmental requirements. The program has also identified problems in their operations that require corrective action, or, more frequently, it has confirmed that environmental requirements were being met.

Health, Safety and Environmental Surveillance Review System

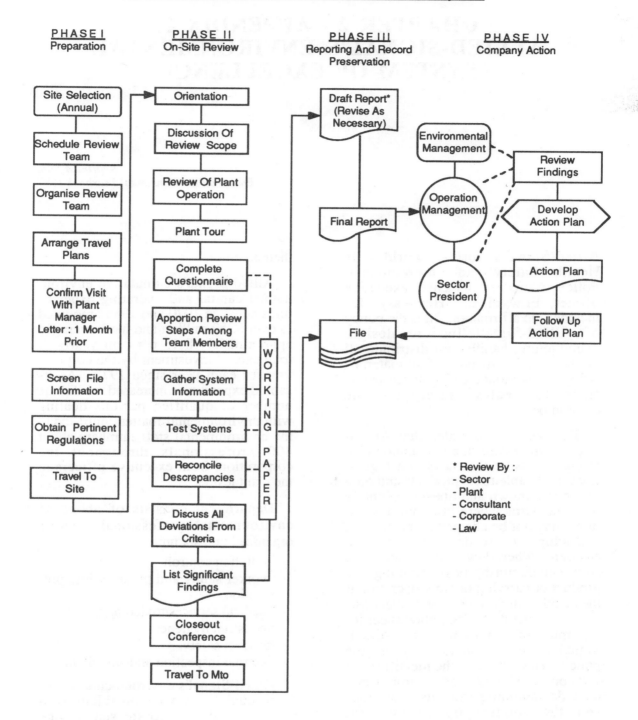

PHASE I
Preparation

PHASE II
On-Site Review

PHASE III
Reporting And Record
Preservation

PHASE IV
Company Action

Site Selection (Annual)

Schedule Review Team

Organise Review Team

Arrange Travel Plans

Confirm Visit With Plant Manager Letter : 1 Month Prior

Screen File Information

Obtain Pertinent Regulations

Travel To Site

Orientation

Discussion Of Review Scope

Review Of Plant Operation

Plant Tour

Complete Questionnaire

Apportion Review Steps Among Team Members

Gather System Information

Test Systems

Reconcile Descrepancies

Discuss All Deviations From Criteria

List Significant Findings

Closeout Conference

Travel To Mto

WORKING PAPER

Draft Report* (Revise As Necessary)

Final Report

File

Environmental Management

Operation Management

Sector President

Review Findings

Develop Action Plan

Action Plan

Follow Up Action Plan

* Review By :
- Sector
- Plant
- Consultant
- Corporate
- Law

CHAPTER 2 - APPENDIX 1:
ALLIED-SIGNAL'S ENVIRONMENTAL SYSTEM OF EXCELLENCE

Reprinted from
European Environmental Yearbook 1987

Allied-Signal's formal world-wide Health, Safety and Environmental Policy, issued by its chief executive officer, Edward L. Hennessy, Jr., includes the affirmation: "it is the policy of Allied-Signal Inc. to design, manufacture, handle and distribute all products and to dispose of all materials safely and without creating unacceptable risks to health, safety or the environment."

The policy also states that Allied-Signal will "establish and maintain programs to assure that laws and regulations applicable to its products and operation are known and obeyed; adopt its own standards where laws and regulations may not be adequately protective, and adopt, where necessary, its own standards where laws do not exist; and stop manufacturing or distributing any product or carrying out any operation if the health, safety or environmental costs are unacceptable". The policy specifies a number of actions to which Allied-Signal is committed to carry out the policy. These include the identification and control of health and environmental hazards stemming from its operations and the sponsorship of accident prevention and occupational health programs to protect employees and the public. The policy also states: "every employee is expected to adhere to the spirit as well as the letter of this policy. Managers have a special obligation to keep informed about health, safety and environmental risks and standards and to advise higher management promptly of any adverse situation which comes to their attention".

Allied-Signal's cumulative environmental capital and operating expenditures between 1970 and 1985 amounted to over $250 million and have averaged approximately 8-15 per cent yearly of total capital investment by the corporation for each year since 1972. As its capital expenditure increased, so did the number of qualified persons dealing with environmental matters until today the environmental staff numbers about 400 professionals throughout the corporation, from executive officers to the plant levels.

Department consists of about 34 environmental professionals and is organized into six units:

- pollution control;
- worker safety and property loss prevention;
- medical services/toxicology;
- industrial hygiene;
- product safety;
- internal inspection and surveillance.

Allied-Signal's environmental effort is headed by a vice president who reports to the corporate senior vice president of technology. In addition to Law Department support, the corporate department works closely with public affairs and others.

Recognizing the benefits of a multi-discipline approach to environmental matters, the directors of the units within the corporate environmental department meet approximately weekly to discuss policy and mutual concerns. The corporate Environmental Affairs Department is responsible for 1) policies, overall programs and coordination, environmental planning, major assurance program and review of environmental financial matters, and 2) training, inspection and follow-up. The operating company, sector and plant environmental staffs (composed of basically the same six functions) have responsibility for the day-to-day hands-on operation.

The success of Allied-Signal's environmental program is perhaps best illustrated by its very successful approach to worker safety and property loss prevention world-wide. Each year the corporate Safety and Loss Prevention group coordinates the setting of safety goals, conducts training programs and upgrades the goals to assure continuing efforts to achieve a solid safety record. The plants are responsible for their safety performance, and safety is taken seriously at every level of the corporation's activities. Allied-Signal is an industry leader in safety performance and the cost of savings go, of course, directly to the bottom line.

Each of Allied-Signal's three operating sectors (Aerospace and Electronics, Engineering Materials and Automotive) have safety staff who are responsible for assisting the line in implementing safety programs and improving safety procedures.

Reports on achievement of safety objectives are made regularly to top management. If the objectives have not been met, steps are taken immediately to rectify the situation. Those involved in Allied-Signal's corporate and operating company management eligible for year-end bonuses for increasing profitability have had a percentage of the bonus based on safety performance as measured against the yearly safety goals established fot that unit.

Allied-Signal strives for an injury reduction rate each year. In its total case incident rate and lost workday cases incident rate both per 100 employees for the past four years was:

Year	Total Cases Incident Rate*	Lost Workday Cases Incident Rate*
1982	2.96	1.26
1983	2.39	0.99
1984	2.03	0.85
1985	1.70	0.66

*US and Canadian incidence rates per 100 full-time employees using 200.000 employee-hours as the equivalent.

Such performance is kept and monitored world-wide by the corporate and Sector Safety Staff, which works with the operations in setting goals, establishing programs, training and follow-up.

Safety engineers employed within the corporation and from outside services periodically inspect Allied-Signal plant locations, checking potential and actual hazards, including process hazards, possible unsafe working procedures and hazardous working areas. After the inspections, the engineers sent letters confirming inspection, and/or recommendations to the plants, which are monitored by management personnel.

During the year, each company and the corporate management not only monitor how well the safety objectives established at the year's beginning are being met, but carry forth a comprehensive safety program tailored to their needs. Allied-Signal's 1986 "Strive for Excellence" safety program revolves around themes:

1) hazard identification/evaluation
2) probable maximum loss
3) government regulations/legal liabilities
4) education and training - communications
5) safety and loss prevention guidelines
6) Accident Prevention Plans & Statistical Objectives
8) Assurance Reviews

These themes are highlighted at management meetings and then programs implemented through plant safety meetings, where employees meet in small groups to permit individual employee participation in problem solving and in executing safety procedures. Each worker is reminded of his responsibility to himself and his fellow workers for maintaining a safe working environment.

At many of the monthly safety meetings, video cassette tapes of about 20 minutes length produced by the corporate safety department at Allied-Signal plant locations are shown. The video tapes are accompanied by workbooks. The same tape may be presented at many locations to develop the current theme of the program. For example, one such tape is entitled "Safety is Caring ... About the Proper Use of Protective Clothing and Equipment". It stresses the role played by protective clothing and equipment in assuring worker safety. Respiratory protection and a variety of gear, both commonplace and exotic, is described. The tape tells employees that injuries due to failure to wear protective clothing are the most preventable of all.

The "Strive for Excellence" safety campaign is publicized in several ways, including display of a series of eye-catching photographic posters and other material distributed at the beginning of the year to plant locations. Programs at different Allied-Signal sites are developed on the same theme concurrently. The themes are changed periodically to accent a different safety concept.

In addition, Allied-Signal conducts special safety training courses for supervisors to teach them how to instruct their workers in safe working procedures. Workshops for all safety professionals are held annually in the U.S., in Europe and in Latin America, for example, at which case histories, current safety programs, and Allied-Signal's comprehensive and strict safety standards are discussed. The Corporation also holds technical loss prevention sessions during the year on topics such as combustion control and fire prevention technology. Some of these programs are on cassettes. One tape, for example, is "the Supervisor as an Educator". It describes the basic educational tools needed by the supervisor in fulfilling his responsibilities in the field of employee safety training.

A second example of the environ-mental program's value is The Risk Assessment Committee (which Allied-Signal calls TRAC) and the Product Integrity Committee (which Allied-Signal calls PIC) to do risk and hazard evaluation and product defect review.

The "Strive for Excellence" safety program and the Product Integrity Committee are part of a fabric on inter-related programs carried out at the company and plant levels "to establish and maintain programs to ensure that laws and regulations applicable to its products and operation are known and obeyed" both in "spirit as well as the letter ...", including "to advise higher management promptly of any adverse situations ...".

Other important programs include:
- Product responsibility, in which the operating businesses plan and carefully carry out their activities in a manner designed to meet their obligations in research, product development, manufacturing, distribution, and sale of its products so as to ensure a high and competitive degree of product safety;
- Comprehension pollution control management, in which the corporation's manufacturing operations are permitted and monitored to carry out proper air, water and solid waste disposal practices, including careful procedures for disposing of wastes in licensed locations;
- Occupational health, medical and nursing guidelines, and education and training, to ensure a high level of attention by company health personnel to the corporation's workers;
- An environmental data system to store and retrieve health and safety, as well as product toxicity and waste management control information, required by regulation and/or critical to the safe operation of the company's manufacturing plants and businesses;
- Assurance Reviews or inspections with follow-ups based on Allied-Signal standards and keyed to Allied programs as well as compliance with local laws and regulations by skilled health, safety and environmental professionals (1) from the corporate group and the operations and

(2) from outside inspections in, for example, the insurance industry. An independent surveillance system which on a sampling basis worldwide, reviews various environmental aspects of the Allied-Signal plant activities and reports to plant management deficiencies found so that they may be expeditiously corrected.

All these programs work together at corporate sector, and plant levels to help Allied-Signal strive toward its goals of environmental excellence.

CHAPTER 2 - APPENDIX 2:
HEALTH, SAFETY AND ENVIRONMENTAL POLICY

It is the policy of Allied-Signal Inc. to design, manufacture and distribute all products and to handle and dispose of all materials safely and without creating unacceptable risks to health, safety or the environment. The corporation will:

- establish and maintain programs to assure that laws and regulations applicable to its products and operations are known and obeyed;
- adopt its own standards where laws or regulations may not be adequately protective, and adopt, where necessary, its own standards where laws do not exist; and
- stop manufacturing or distributing any product or carrying out any operation if the health, safety or environmental risks or costs are unacceptable.

To carry out this policy, the corporation will:

1. Identify and control public health, safety and environmental hazards stemming from its operations and products;
2. Conduct accident prevention, product safety and integrity, occupational health and pollution control programs to safeguard employees and the public from injuries or health hazards, to protect the corporation's assets and continuity of operations and to protect the environment;
3. Conduct and support appropriate research about the health, safety and environmental effects of materials and products handled and sold by the corporation and share promptly any significant findings with others, such as employees, suppliers, customers, government agencies or the scientific community; and
4. Work constructively with trade associations, government agencies and others to develop equitable and realistic laws, regulations and standards to protect public health, safety and the environment.

Every employee is expected to adhere to the spirit as well as the letter of this policy. Managers have a special obligation to keep informed about health, safety and environmental risks and standards and to advise higher management promptly of any adverse situation which comes to their attention.

CHAPTER 2 - APPENDIX 3:
RISK OF 'SMOKING GUN' PAPERS IS OUTWEIGHED BY THE BENEFITS

ADD TO:

FIRST ARTICLE SUITABLE PART]

Reprinted from
Preventive law reporter

By Thomas D.KENT
Staff Vice President and Associate General Counsel
Allied-Signal Inc., Morristown, New Jersey

The Uninspected Inevitably Deteriorates"
- Dwight D. Eisenhower

Those words by a former five-star general and president are very appropriate to express the central theme of the environmental compliance programs at Allied-Signal.

This article will focus primarily on Allied-Signal's (1) corporate-wide Environmental Surveillance Program (2) Environmental Assurance Reviews, and (3) its Annual Environmental Assurance Letter Program. These programs might not have materialized, however, if the corporation had not first obtained top management support, as demonstrated by its strong environmental policy.

The author has had Allied-Signal or predecessor companies as his principal client for over 23 years. For most of that time, the corporation was known as Allied Chemical Corporation. Except for the Annual Environmental Assurance Letter program, most of what will be described had its genesis under Allied Chemical.

Some of the programs were undoubtedly created, at least in part, in response to the Kepone incident in the early and mid-seventies. The facts of that incident are not relevant insofar as this article is concerned, but what happened afterwards does have some relevance since corporations, like people, tend to learn from, or react to, their own experiences.

The corporation was a target of an unfriendly CBS "60 Minutes" segment, a Congressional investigation, a grand jury investigation followed by indictments and a criminal trial, numerous private damage suits, two stockholder derivative actions, an SEC complaint and Consent Order and, finally, a qualified auditor's opinion based in part on the foregoing.

The ill-wind Kepone, however, may have blown some good. The positive response by Allied Chemical, as demonstrated by its strong environmental policy and programs, put the corporation in the forefront of environmental compliance so that it eventually earned a national reputation as an environmental "White Hat". Hopefully others can learn from the experiences of Allied Chemical and need not live through traumatic events to adopt suitable environmental compliance programs.

The Corporation's Health, Safety and Environmental Policy is fairly succinct. The key elements of the Corporation's commitment are that it will:

1. "... design, manufacture and distribute all products and to handle and dispose of all materials safely and without creating unacceptable risks

to health, safety or the environment";

2. "Establish and maintain programs to assure that laws and regulations ... are known and obeyed";

3. Adopt *own standards* where laws or regulations may not be adequately protective;

4. "Every employee is expected to adhere to the spirit as well as the letter of this policy. Managers have a special obligation to keep informed ... and to advise higher management promptly of any adverse situation ..." (emphasis added).

Other programs such as TRAC/PIC, our waste site inspection and waste minimization programs, Strive for Excellence Training and Information Guides, video training tapes, training workshops and seminars, legal compliance booklets, as well as specific manuals and guidelines covering subjects such as Occupational Health, TSCA, Product Responsibility, Hazardous Materials Management, Hazardous Waste Generation and Pollution Control all are integral components of environmental compliance.

TRAC/PIC is an acronym for the Risk Assessment Committee and Product Integrity Committee. TRAC was created in 1977 to enable the corporation to more effectively comply with TSCA § 8 (e) "Substantial Risk" reporting requirements. It soon encompassed Consumer Product Safety Act 12 § 5 (b) "Substantial Product Hazard" reporting requirements and those under the Food Drug and Cosmetic Act and Federal Insecticide, Fungicide an Rodenticide Act.

After Allied's acquisition of Bendix, TRAC was combined with their Product Integrity Committee (PIC) which was focused on other statutory reporting requirements, such as those covered by NHTSA or FAA. Both committees have senior representatives from appropriate disciplines (e.g. Pollution Control, Product Safety, Industrial Hygiene, Medical Toxicology, Quality Assurance, Engineering, Insurance, Law).

Problems "bubble-up" from plant-level coordinators to the committees where they are assessed and, if appropriate,

reported to the statutorily specified government office (e.g., EPA's Office of Toxic Substances). Results of all items considered flow back to those who originally surfaced the concern. Since inception, TRAC has focussed on 137 items, 21 of which were reported to a government office.

But in my view, the central core of the Allied-Signal program stems from its Environmental Surveillance, Environmental Assurance and Annual Environmental Assurance Letter programs.

It may come as a mild shock to the lawyer reader of this article, but the real secret to compliance in this area is not the lawyer-he or she is merely one cog in the compliance wheel. Rather, the secret lies in the active involvement by a sufficient number of competent multidisciplined environmental professionals. These folks should be primarily at the plant level, but with a sufficient number on staff to develop, implement and monitor the various environmental programs.

> The secret lies in the active involvement by a sufficient number of competent multidisciplined environmental professionals.

ENVIRONMENTAL SURVEILLANCE PROGRAM

We deliberately avoided use of the term "audit", although we recognize that term to be uniquitous. Even the EPA has issued an interim guidance entitled "Environmental Auditing Policy Statement" 50 Fed. Reg. 46504 (11/8/85).

In a financial context the term "audit" is accepted and understood. Environmental "auditing", on the other hand, is a relatively embryonic science. There are no equivalents of the Certified Public Accountant (CPA), Financial Accounting Standards Board (FASB), generally accepted auditing standards (GAAS) or generally accepted accounting principles

(GAAP). Furthermore, there is a substantial body of case law analyzing and assessing the responsibilities of auditors.

Allied-Signal's corporate headquarters is in New Jersey. In 1983 the Supreme Court of that state issued a unanimous decision which greatly expands the class of those who can assert negligence claims against certifying accountants. *H. Rosenblum, INC. v. Adler*, 93 NJ 324 (Sup. Ct. 12983).

Some observers believe that the tone and tenor of the *Rosemblum* opinion are such as to make it highly likely that the principle of the case will be expanded to cover claims by all reasonably foreseeable "consumers" of negligently audited financial statements and, furthermore, that the privity doctrine espoused in the seminal case of *Ultramares v. Touch*, 255 NY 170 (1931) (Cardoza, Ch.J.) will fall in an increasing number of jurisdictions. Whether that will actually be the case, I do not pretend to know.

Privity, or its practical equivalent, still exists in New York, at least, as the predicate for accountants' liability in negligence. See *Credit Alliance Corp. v. Arthur Andersen & Co*, 65 NY 2d 536 (1985). But for us, the answer was clear. Why even use a term ("audit") which has a clear meaning with potential third-party liabilities in one context? Don't even give a future trier of fact the opportunity to draw the analogy - just avoid the term.

This Environmental Surveillance Program was adopted in 1978, after a gestation period of over a year. Its objective is to provide top management and the Board of Directors with an independent verification that our operations are in compliance with Allied-Signal's health, safety and environmental (HSE) policy and procedures as well as HSE laws and regulations.

The program is headed by the Corporate Director, Surveillance, a senior member of the Corporate HSE Department. It provides a historical snapshot showing through factual observations where a facility stands with respect to current laws/regulations and corporate policy and procedures. Deficiencies are noted for follow-up and correction, but it is not the function of the Surveillance Review to make recommendations to change or modify programs to possibly improve the overall programs from a future risk-management perspective. The latter is the primary function of the Environmental Assurance Reviews.

The Surveillance Program was designed by the Surveillance Director in conjunction with the Law Department our internal financial auditors (to pick up, where deemed applicable, the "disciplines" of generally accepted auditing procedures) and our outside consultants (Arthur D. Little).

Surveillance Reviews cover six HSE disciplines or areas; water pollution control and spill prevention, air pollution control, solid and hazardous waste, safety and loss prevention, occupational health and medical, product safety and integrity.

In 1986, approximately 50 manufacturing facilities were reviewed-out of 300-plus worldwide. The Surveillance Director, the various corporate HSE directors, and A. D. Little select the facilities to be surveilled from a pool of plants that the group collectively believes might pose significant environmental risks. Random number generators are used to exclude personal biases, and an attempt is also made to balance HSE functional areas as well as major business areas.

Each facility is given substantial advance notice of a pending surveillance review. In addition, as part of its advance preparation; it receives a copy of the then - applicable review protocols. We are not policemen; the reviews are intended to be constructive and, at least for the most part, are perceived by the reviewees in that light.

Depending on the size and nature of the facility reviewed, the surveillance team will consist of 2 to 4 environmental professionals and take 3 to 4 days to complete. One inspector is always the Corporate Surveillance Director or a member of his staff; another is always from A. D. Little with the remainder, if any, from the corporate staff with expertise in the discipline being reviewed. In addition, there is often an observer from the operating unit being reviewed.

It has been our experience that it is more effective, at least in the U.S. and Canada, to review only one discipline per visit-overseas, because of travel time and costs, multiple disciplines are sometimes reviewed during a single visit, although this tends to place an extra burden on the plant.

At the conclusion of the review, a complete detailed oral exit interview (along with written lists of findings) is held with the plant manager. In addition, prior to issuance of the final report, the plant manager gets an opportunity to comment on a draft report. Misunderstandings, if any, are usually cleared up at the exit interview or in response to the draft report.

Shortly thereafter a formal written report goes to the company President with a request for a written reply indicating corrective action taken or anticipated in response to each finding or deficiency noted. Having the President in the loop helps ensure that sufficient funds will be appropriated to timely accomplish the corrective action.

> We are not policemen; the reviews are intended to be constructive and ... are perceived by the reviewees in that light

As a further step to provide verification of internal compliance efforts, roughly 20 per cent of these action plans are subjected to review for completion status by the surveillance group. These follow-up reviews are conducted 1 to 2 years after the initial review.

These formal reports are *always* written by the Surveillance Director or a member of his staff. After indicating the scope of the review, they set forth

factual observations (no conclusions) measured against (a) federal - state - local regulatory standards, (b) corporate policies and procedures, and (c) local control systems.

Twice a year an overview report as to results of the environmental surveillance reviews (as well as other environmental programs) is made to the Corporate Responsibility Committee of the Board of Directors. An A. D. Little representative is present and, at the conclusion of the meeting, inside personnel (including the CEO) are excused and the outside directors have a chance to privately cheat with the ADL representative.

Some people are fearful that the reports generated by reviews such as I have described will inevitably create a "smoking gun" and prove too dangerous in the long run. And there is no question that there are concerns associated with the program.

Obviously, to accomplish the purpose, these reports must be accurate and complete; they must communicate in a meaningful way. Since they highlight deficiencies, if left unattended in the files they could indeed be a "smoking gun" that shoots.

In addition, there are review team-work papers which also could be a potent weapon in unfriendly hands. Some undoubtedly document violations and thus would make an enforcement case for the government, or even a citizen group. They would tend to provide "knowledge" to corporate officers and thus could open the door to possible criminal actions (knowing or willful violations, etc.) if the deficiencies noted are not promptly addressed. The same paper flow might support punitive damage claims as well in private suits. Furthermore, they might reveal trade secrets as to production processes, provide ammunition to the SEC in enforcement actions (failure to publicly disclose material adverse environmental uncertainties, etc.); provide information to someone trying to block an acquisition or merger (as I recall, it was the "scorched earth" defense of a target company to a proposed Occidental petroleum/Hooker takeover that unleased Love Canal), etc.

But in my judgment the benefits far outweigh the risks, especially if one recognizes the risks and takes steps to minimize those risks. We will address the benefits later; first let us examine the steps we take to minimize those risks. Initially, we take great care in drafting the reports-facts, not conclusions. They are very structured, always written by a small group of professionals after much training.

We also have a rigidly adhered-to document retention program. The formal reports and corrective action plans are retained 10 years (CERCLA and RCRA regulations may require 50-year retention in some instances). There is only one set of working papers — retained three years- and all rough notes and draft reports are supposed to be destroyed as soon as the action plan in response to the formal report is received.

One step which might minimize some risk that these papers could be used against us that we deliberately did not take, however, was to run the program through the Law Department to impart "attorney work product" or "attorney-client privilege". In 95-plus per cent of the cases, no legal advice is really being sought — deficiencies and weaknesses measured against the law/regulations, corporate policies and procedures are spotted, and the primary goal is to fix those deficiencies promptly, not prepare for litigation. If the whole program were run through the Law Department, we might not only prove to be a bottleneck, but even of greater concern, lose the very protection we sought when it really counts by overreaching or abusing the privilege.

ENVIRONMENTAL ASSURANCE REVIEWS

These reviews are thorough, in-depth inspections by the professionals in the four basic HSE disciplines (pollution control, safety and loss prevention, product safety and integrity, occupational health and medicine) which make up the core group of professionals in the Corporate HSE Department.

As noted above, these reviews are more programmatic, utilizing risk assessment techniques and looking to the future in an ever-continuing effort to improve our procedures and control systems. The ouput of an assurance review will ultimately be an action plan to implement the agreed-upon corrective steps.

There is perhaps a greater risk here of inadvertent creation of "smoking gun" papers since more people are involved and there is a less rigid document creation/retention program. Nevertheless, we are still dealing with a manageable group of professionals, all of whom have a close personal working relationship with the Law Department and have been reasonably well sensitized to the potential dangers of report writing.

Although not typical, if in the course of an assurance review the inspector believes he/she has found "a bomb waiting to explode" which might require an external report to an agency or could otherwise trigger administrative or legal proceedings, that review report will not only be edited by a lawyer, but will be addressed to the general counsel of the operating sector in anticipation of litigation and will in fact seek legal advice. This does not happen often, but it has happened frequently enough (especially when reviewing newly acquired facilities that have not long enjoyed the environmental culture of Allied-Signal) so that I am confident that there are few, if any, loose written cannons on the deck and that not too much is falling between the cracks. Last year we conducted about 150 of these environmental assurances reviews.

Not technically falling within our definition of assurance reviews, but worthy of brief comment at this point, are the 1.400 or so external inspections of our operating locations in the U.S. and Canada carried out by the engineering affiliates of our insurance carriers. The protocols, policies and procedures used by these many independent sets of eyes are those of Allied-Signal, not the less stringent carrier procedures generally used, and relate to (a) Boiler and

Machinery, (b) Property Loss, and (c) Personnel Safety. Needless to say, these inspections also generate numerous recommendations which have to be implemented or explained away — never left unaddressed.

ANNUAL ENVIRONMENTAL ASSURANCE LETTER PROGRAM

To my way of thinking, this is perhaps the most effective program of all, in that it vividly demonstrates top-managements support for environmental compliance and serves to raise dramatically the level of environmental consciousness down to the corporate grass roots. It helps tie up the loose ends and create a neat package.

Each June the CEO formally requests each operating Sector President to go on the line that he/she understands the corporate HSE policy (remember — *spirit* as well as letter) and has taken steps to bring that policy to the attention of employees in that sector. He/she goes on the line to state — or explain why not — that control systems are in place to assure compliance or adherence to the law and regulations as well as corporate HSE policy and procedures; that the sector has satisfied its responsibilities with respect to permits, notifications and other actions required by law; that significant violations/deviations have been reported to the corporate HSE Department; and that he/she knows of no matters which could have a materially adverse future impact on any of the sector's business areas.

This last is key — in a company the size of Allied-Signal, few environmental problems would have a material adverse impact in the Securities and Exchange Commission sense. By breaking it down to separate business areas, the CEO's letter helps to ensure that problem areas on a smaller scale "bubble-up" and thus can be promptly corrected or otherwise handled.

Now a Sector President is not going to go on the line in writing to the CEO in the manner described without substantial written back-up. And that's what happens. Executives right down the line to the plant manager and beyond execute similar letters, and those flow back up the line with action plans and schedules, with target compliance dates, until the Sector President has what he/she deems adequate to enable him/her to sign their own environmental assurance letter to the CEO. That's why the process takes about 3.5 months. Like the Surveillance Reviews, summaries of these Environmental Assurance Letters are reviewed with the Corporate Responsibility Committee of the Board.

ENVIRONMENTAL EXCELLENCE EQUALS DOLLARS IN THE EXECUTIVE'S POCKET

The programs described demonstrate a corporate-wide commitment to environmental excellence, but Allied-Signal, like most corporations, has a variety of commitments (e.g., shareholder profit) which to some executives may not always seem compatible. Our people are no different from others; they are human and for the most part can be relied upon to do what is "right", especially if what is right is financially rewarding.

All key executives develop annual action plans which detail steps necessary to accomplish selected objectives. Some goals are financial, but about 40 per cent of an executive's incentive compensation is dependent on meeting non-financial objectives. A non-financial goal for many operating executives is environmental and safety program compliance.

Furthermore, since litigation cost (including internal lawyer time), fines, penalties, settlements and remediation costs are charged back to the business unit, the failure to sail an environmentally sound ship is likely to cause an executive's financial objectives to founder as well.

BENEFITS OF ASSURANCE/SURVEILLANCE PROGRAMS FAR OUTWEIGH RISKS

The Allied-Signal Automotive Sector has a well-known TV advertisement for its FRAM filters which depicts a mechanic saying, "You can pay me now or pay me later". In print and point-of-purchase ads the phrase, "Pay a little now or a lot later" is used. The point of these ads is equally applicable to environmental compliance.

For the corporation, the monetary benefits cannot be accurately quantified, but are, nevertheless, real and substantial. Furthermore, the risk of criminal prosecution is greatly reduced, as are the chances of a private plantiff being awarded punitive damages.

Good corporate citizenship in this also means a safer place to work, a safer environment for neighbours and customers, as well as a cleaner environment generally. It helps to improve employee morale, instill pride in one's employer, which in turn can mean higher productivity.

Insofar as the individual is concerned, he/she is also less apt to be criminally prosecuted. In this regard, consider *U.S. v. Park*, 421 U.S. 658 (1975) where the Supreme Court affirmed a lower court finding that the president of a billion-dollar company may be guilty of criminal behaviour (violation of the Food Drug and Cosmectic Act) "by reason of his position in the Corporation, responsibility and authority either to prevent in the first instance, or promptly to correct, the violation complained of, and that he failed to do so". (*Id.* at 673-4).

Four Warner-Lambert executives (including the corporate director of safety) were indicted, following an explosion in 1976 which killed six employees, for criminally negligent homicide and reckless manslaughter. Although the N.Y. Court of Appeals eventually dismissed the indictments. *People v. Warner-Lambert Co.*, 51 NY 2d 295 (1980) cert. den. *N.Y. v. Warner Lambert*, 450 U.S. 1031, you can be

sure it was a living hell for the individuals involved.

More recently, top officers of Film Recovery Systems, Inc. were convicted of murder (along with 14 counts of reckless misconduct) and sentenced to 25-year prison terms as a result of the death of a worker (cyanide poisoning). *People v. O'Neil,* et al. reported in *The New York Times* (7/2/85, Sec. A, p. 11, col. 4). These convictions are on appeal before the Illinois Appolate Court, oral arguments having been held July 7, 1987.

Although more dramatic, the *O'Neil* case parallels the *Magnet Wire Corp.* case which involved indictments against the corporation and five corporate officials charged with aggravated battery and reckless conduct for, among other things, failing to provide necessary safety training and equipment, and operating a plant under dangerous conditions in which employees worked with toxic chemicals. On June 29, 1987, the Illinois appellate court found the Congress' intent in passing OSHA in 1970 was to preempt application of criminal statutes by states to conditions in workplaces that were specifically regulated by the Act. According to the Bureau of National Affairs' *Chemical Regulation Reporter* (July 10, 1987, p.690) the Cook County Assistant States' Attorney has filed notice of its intention to appeal this preemption ruling. In any event, win or lose, no company or individual wants to be caught up in these types of proceedings.

In short, environmental excellence pays in many ways. The proactive programs described may not be appropriate for all manufacturing companies, but they work pretty well for Allied-Signal. Furthermore, these programs are not static, we are continually evolving new wrinkles which we believe make them even more effective.

ALLIED - SIGNAL
ENVIRONMENTAL ASSURANCE PROGRAM

· ANNUAL HEALTH, SAFETY AND ENVIRONMENTAL ACTION PLAN
· ASSURANCE LETTER PROCESS
· SAFETY PERFORMANCE ACCOUNTABILITY
· TRAC - CHEMICAL CONTROL
· ASSURANCE INSPECTION PROCESS
· EMERGENCY RESPONSE PLAN
· COMMUNICATION COMMITTEES
· DISCLOSURE PROGRAMS - BUBBLE UP
· SURVEILLANCE AUDIT

OBJECTIVE

The surveillance programme was initiated at the request of the Board of Directors to provide top management and the Board with an independent verification that the Corporation's operations are in compliance with applicable law and health, safety and environmental policy and procedures

APPROACH

· Employ Internal/Financial Audit Procedures
· Conducted by a Team
· Limited to Consideration of Hazards for which Criteria have been Established
· Focus on Deficiencies
· Directed at Material Exposure

REVIEW SITE/SUBJECT SELECTION

· Pool - Locations With H, S or E Significance
· Random Selection
· Cross Section
    - H, S & E Disciplines
    - Business Interests

REPORTING

Who	When	What
Plant Health Safety & Environmental Supervision	When Noted	All Deficiencies Noted
Location Executive	Periodic & Exit Meeting	All Significant Deficiencies Noted
Health Safety & Environmental Staff	Informal & Formal Report	The Most Significant Matter
Sector President	Formal Report & Conference	The Most Significant Matter
Board Corporate Responsibility Committee	Regular Meeting	General Status

CHAPTER 3 - THE APPROACH TO ENVIRONMENTAL AUDITING IN THE ROYAL DUTCH/SHELL GROUP OF COMPANIES

I.J. GRAHAM-BRYCE
Shell Internationale Petroleum

ROYAL SHELL PLC.

1. INTRODUCTION

Environmental auditing, as outlined below, is regarded as an important activity for the Royal Dutch/Shell Group of Companies. It is seen as part of a company's own responsibility and as just one element in a coordinated on-going approach to environmental protection. The underlying philosophy is incorporated in the Statement of General Business Principles for the Royal Dutch/Shell Group which includes the declaration:

"It is the policy of Shell companies to conduct their activities in such a way as to take foremost account of the health and safety of their employees and of other persons, and to give proper regard to the conservation of the environment. In implementing this policy Shell companies not only comply with the requirements of the relevant legislation but promote in an appropriate manner measures for the protection of health, safety and the environment for all who may be affected directly and indirectly by their activities".

These principles are amplified in Guidelines on Health, Safety and Environmental Conservation which indicate that Shell Companies, amongst other requirements, apply the best practicable means to preserve air, water, soil and plant and animal life from adverse effects of their operations and to minimize any nuisance that may arise. The above statements are being supplemented by guidance documents on environmental management which provide practical advice on how the objectives may be achieved — a structured and comprehensive scheme intended to prevent unacceptable effects in the environment from products and operations. All stages are considered from initial conception and planning to final termination and decommissioning. Environmental auditing is considered to be one of the key procedures to be applied during the operational phase of activities.

2. THE ROYAL DUTCH/SHELL GROUP

In any description of environmental auditing within Shell Companies, it is necessary to convey the size and complexity of the enterprises involved.

The Royal Dutch/Shell Group of Companies are active in four main business sectors: Oil and Gas, Chemicals, Metals and Coal. Total sales are approximately $100 million per annum.

The main business is conducted by operating companies of which there are several hundred in over 100 countries. The activities of the operating companies are supported by Service Companies located in the Netherlands and the UK which provide specialized advice and services, including expert advice on environmental matters.

It will be apparent that the requirements for environmental auditing in the units of such a large, multifaceted group of companies will vary in detail and a complete description is not possible. However the principles and the general approach are common to all situations and will be described here.

3. THE DEVELOPMENT OF ENVIRONMENTAL AUDITING WITHIN THE ROYAL DUTCH/SHELL GROUP

Although the term "environmental audit" is currently being used, it is important to emphasize that the essential concepts are far from new. Various forms of environmental performance appraisal have been practised within Shell Companies for many years as one component of environmental management. These have become more systematic and standardized and their value as a management tool has been increasingly recognized.

To quote a few examples from across the Group: Health, Safety and Environment (HSE) Audits for Agrochemicals formulation and repacking plants have been conducted on a routine basis world-wide for several years; environmental audits of oil and gas exploration and production operations take place on a three-year cycle and HSE Audits are regularly held in the Chemicals Manufacturing sector. The nature of the facility being audited may vary widely from small activities (e.g. a single gas terminal, a lube oil warehouse, a formulation plant, a service station) to a very large scale operation (e.g. the activities of a large company involved in exploration and production of oil and gas, a petrochemical complex, a refinery). Correspondingly, within the basic principles environmental audits differ in scope and type of facility being audited.

Individual Shell Companies may conduct internal audits, using local staff, relatively frequently (e.g. annually); these are complemented by more extensive audits, involving staff from other locations and from Service Companies, taking place at longer intervals, typically three to five years.

4. DEFINITION OF ENVIRONMENTAL AUDITS AND POTENTIAL BENEFITS

There is no universally agreed definition and system of environmental audits and many different terms have been used for the concept. Environmental Auditing as practised within Shell Companies corresponds broadly with the definition adopted by the International Chamber of Commerce (ICC). If these purposes are to be fulfilled with full cooperation of those involved, it is considered essential that auditing should be seen as the voluntary responsibility of the company itself, with the audit report being provided to the company management. The major value to the process is as a management tool which provides timely information on environmental performance in relation to goals and intentions.

In stating this important point of principle, it is fully accepted that industrial activities should be subject to regulations and permits intended to limit their impact on the environment, for example consent levels for effluent discharges, emissions to air and waste disposal.

Monitoring compliance with these regulations is an important continuing activity. Environmental audits, however, go further and include a comprehensive review of policies, managements systems, organization and practice and compliance with internal standards.

As already made clear, the primary benefits of environmental auditing are to indicate in good time whether environmental measures are satisfactory and to assist with and substantiate compliance with company policy, laws and regulations. A related advantage for companies is, of course, reduced exposure to litigation and regulatory risk. Experience within Shell Companies has demonstrated that environmental audits can have additional benefits, for example:

- increasing awareness of environmental policies and responsibilities;
- providing an opportunity for

management to give credit for good environmental performance;

- identifying potential cost savings, for example those which might arise from waste minimization;
- providing an up-to-date environmental data base, which can be drawn on when making decisions in relation to plant modifications etc. or for use in emergencies;
- evaluating training programmes and providing information for use in training staff.

5. GENERAL APPROACH TO ENVIRONMENTAL AUDITING WITHIN THE ROYAL DUTCH/SHELL GROUP

The three essential steps in environmental auditing are:

- the collection of information;
- the evaluation of information collected;
- the formulation of conclusions, including identification of aspects needing improvement.

The aspects covered include policy and planning, organization for environmental matters, awareness and training, performance of and requirements for control equipment, monitoring procedures, liaison and communication within the company and externally, and any effects on the environment around the facility.

Practical experience has indicated that several factors are most important if full benefit is to be gained from environmental auditing. For example the audit team must be carefully selected to ensure objectivity and professional competence. For the larger audits conducted within Shell Companies, it has been found generally beneficial to include representation from the location being audited, but it is essential to ensure that overall the team is sufficiently detached to provide an independent view. It is also important to have well defined, systematic procedures which are known and understood by all concerned. Staff should be informed beforehand about the purpose and scope of the audit. It

will be self-evident that the audit should be properly documented in written reports, concentrating on factual objective observations and submitted in good time.

Any appraisal of performance is best achieved by objective quantitative measurement and comparison with reference standards. At first sight this might appear difficult in the case of environmental performance, but various yardsticks may be suggested, for example quantities and nature of wastes produced, emissions and effluents levels, numbers of staff receiving environmental training etc.

An important consideration within a large, complex Group of Companies such as the Royal Dutch/Shell Group is that of comparability and consistency of approach. Various steps can be taken to foster this, for example the adoption, as far as possible, of standard procedures, the interchange of audit team members and the inclusion of experts with wide experience.

Finally the value of an environmental audit depends strongly on active follow-up of points raised and recommendations produced. This can be facilitated by developing an environmental action plan which addresses relevant issues.

6. OUTLINE OF ENVIRONMENTAL AUDIT PROCEDURES

The need for flexibility in relating the specific details of an audit to the particular circumstances has already been stressed. Nevertheless most audits follow the following general sequence.

6.1 Preparatory activities

Once a company management has decided on the need for an environmental audit, a team leader is identified. The terms of reference and scope of the audit are defined in discussion with the team leader. An audit team, appropriate for the objectives of the specific audit, is then formed.

Typically, for the larger audits, the team would be drawn from different departments within The Royal Dutch/Shell Group to provide expertise in general environmental matters (policy, regulations, environmental management), specific environmental expertise (ecology, environmental toxicology, fate and behaviour of potential contaminants), abatement technologies and operational aspects. The number of team members will obviously vary, depending on the complexity and size of the facility being audited, but is commonly between four and eight persons.

The next step is to assemble a pre-visit information package. Usually the site would provide background information on the nature of the operations and environmental considerations relevant to the audit exercise. The completed information package should preferably be in the hands of the audit team some weeks before the site visit. There is no doubt that the quality of the preliminary information can greatly influence the efficiency of work by the audit team.

On the basis of the information available the audit team identifies the main areas for consideration, develops a visit programme and allocates specific tasks to team members.

At the location to be audited, the management informs all employees in good time that an audit will take place and provides background information on objectives.

6.2 - On-site visit

Depending on the scope of the audit, the size and complexity of the operation and the number of team members, the on-site visit may take between one day and several weeks. The key elements of the visit are inspection of the facilities and interviews with staff.

The visit generally starts with a "kick-off" meeting between the senior management of the site and the audit team to re-confirm objectives and procedures and settle any outstanding questions. According to the agreed audit programme, the team will then visit the necessary operational sites, facilities and surrounding neighbourhood and conduct interviews with staff.

The selection of staff for interview should achieve a good cross sectional impression of the operations and procedures at the site. The interviews include senior, middle and junior staff of relevant departments. It has been found helpful to conduct interviews in accordance with a protocol agreed beforehand, using a flexible checklist.

It is considered important that the audit team prepares its draft report during the visit and presents the findings and any recommendations for areas requiring attention in a final "wrap-up" discussion with senior management of the company before leaving.

6.3 - Post-visit activities

Within a short time (typically one month) the audit team produces a full report for comments by the site management. The report is then finalized as soon as possible. On the basis of this report the site management (assisted as appropriate by the audit team) develops the plan to address the findings. To maintain momentum and optimize the benefits from the audit, this should be done as promptly as possible.

7. CONCLUSIONS

The concept of environmental auditing is still evolving and developing, but experience within the Royal Dutch/Shell Group has demonstrated its value. It is increasingly accepted and welcomed and although there may be animated discussions about audit findings and recommendations, in practice there are few basic disagreements.

The widespread interest in environmental auditing within and outside industry is evidenced by this publication. It is to be hoped that as auditing develops and extends, it continues to be recognized that the full value from this management tool can only be obtained if it remains part of industry's internal procedures in meeting their responsibilities.

CHAPTER 4 - ENVIRONMENTAL AUDITING IN THE BP GROUP

Eric B. COWELL
British Petroleum International Limited

1. INTRODUCTION

The BP Group is one of the world's major international oil and natural resources groups comprising some 1.900 companies, employing some 126.700 people world-wide and with activities and interests in more than 70 countries. The BP business includes Exploration for Oil and Gas, Refining and Marketing, Chemicals, Coal, Minerals, Shipping, Nutrition. The whole Group operates under a corporate environmental policy with amplifying principles and supporting guidelines. The issue of general policy is comparatively easy; the real challenge lies in devising management procedures to ensure that policy is actually implemented at all levels of company activities. To reassure the Group Head Office that policy is adequately translated into operating procedures to protect the environment, requires some form of monitoring or checking system. Such checks in the BP Group have become known as environmental audits.

In this paper we discuss the term "environmental audit" only as it is applied within the BP Group of companies.

Environmental auditing has been conducted within the BP Group from at least the early 1970s. The word audit was first used in this context by the company in 1972, when a refinery and petrochemical complex was subjected to a stringent examination by a team from head office chaired by the General Manager of what was then called the Environmental Control Centre. Subsequently similar activities were called environmental reviews. These in depth examinations included refineries, ocean terminals etc. The formal adoption of the term "environmental audit" was in 1985, following the terminology commonly used in industrial safety reviews.

2. OBJECTIVES

The objectives of environmental auditing are:
- To ensure that cost effective systems of environmental protection management are in use at sites;
- To ensure that the standard of environmental protection is sufficient to meet current regulatory constraints and that systems are in place to cope with future regulatory demands and BP policy requirements;
- To ensure that the standard of environmental protection management is sufficient to promote good relations with local, national and international communities and to promote the image of the BP Group;
- To minimize actual or potential liability;
- To promote environmental awareness at the facility;
- To provide reassurance and make recommendations to management on the above.

3. IMPLEMENTATION OF BP GROUP POLICY

Within the BP group the chief executive of each business is required to ensure that there is an environmental policy, tailored to the needs of this business. Appropriate derivative statements enable staff to understand their roles and responsibilities. Each operating unit and location will in turn have their site specific policy and procedures. It is therefore inevitable that between the corporate head office and the often remote locations in which operations may be conducted, policy can be misunderstood or its importance perceptually diminished. In practice, to strengthen policy and to ensure that policy is implemented requires a number of different but complementary audit approaches. Responsibility for auditing may be split to meet the needs of the group's head office and the needs of the businesses.

4. THE CORPORATE AUDIT

A Corporate Environmental Audit is essentially authorized by the main board of the parent company. Corporate Audits themselves fall into a number of categories. Typically the unit audited is a BP business, e.g. Chemicals International. It is chiefly concerned with the organizational structure to ensure that roles and responsibilities are understood by chief executives, and to examine the organisation structure that purports to deal with EPM: line management responsibilities, technical and advisory support, vertical and lateral communications, etc.

The Audit Team consists of a leader, an audit secretary and a representative of the business concerned. An initial interview programme is agreed and arranged.

In some cases it may be possible to compare the results of the audit with those of audits previously carried out, or with the recommendations and conclusions of environmental impact assessments. These documents should be obtained before the audit begins.

In a typical Corporate Audit of a business the first interview is with the chief executive. After an introduction to the audit to emphasize its authority, the questions relate to policy, understanding of it and the chief executive's views on how his organization to implement it is established and its effectiveness. Supplementary questions deal with communications, awareness, issues,

relationships to the other BP businesses. Many of the questions probe the attitudes to environment of importance to the business, its staff, its customers and the public.

Subsequent interviews are with the line managers and they will be investigated right down to on site visits at selected sample locations.

Site visit locations are selected from information gained during the initial interviews at the business head office. While they are primarily still concerned with organization, matters of policy implementation, awareness and communication channels, occasionally site visits require a certain degree of "nuts and bolts" investigation. In this way the strength and sense of urgency conveyed by management to the workforce and the upwards and downwards efficiency of communications becomes apparent.

This type of organization audit, specifically to examine the efficiency and effectiveness of management in implementing corporate environmental policy when translated into the operating area, is possibly unique to BP.

5. ISSUES AUDITS

A second type of Corporate Audit which is just being initiated relates to the auditing of how the group deals with specific environmental issues of key concern. Several BP group businesses have operations which are by their geographical location potentially harmful to TRF. These vary from exploration for oil, gas, coal, minerals to marketing operations.

Issue audits therefore involve an evaluation of policy, operating procedures and other guidelines set against actual operating practice within all businesses. There are many issues which merit such audits. They are important to reassure a concerned corporate head office that the operating businesses are themselves concerned and reponsible in their approach. In addition, since most environmental protest groups direct their efforts at the corporate image, it is essential that remote sections of the group in, for example, third world countries meet principles and practice which fit group corporate policy so that the head office is not left in an exposed position.

6. ACTIVITY AUDITS

A third form of corporate audit is to evaluate the implementation of corporate policy in activities which cross business boundaries. For example, in 1987 we conducted an environmental audit of all shipping operations in the BP group. This was an audit not only of our tanker operations but also of all vessels including barges, rig supply vessels, gas ships etc. It was directed at environmental policy for the operation of vessels and the organizational structure for ensuring its implementation. This audit included sample interviews with ship crews, jetty and deck crews, examination of ship operating instruction and procedures and the environmental requirements of chartered ships and vessels operated by contractors.

All corporate audits are conducted in a cooperative manner and the aim is not to police and punish but to investigate and assist. All audits reveal to some degree or other organizational weaknesses which require correction. These can arise for a number of reasons which include:

- Failure of management to understand the strength and intent of corporate policy;
- Lack of environmental awareness. Usually attributable to poor communications, insufficient knowledge, or the overall complexity of the issue;
- Changes of personnel and inadequate job descriptions which fail to pin-point environmental responsibility;
- Organization inadequacy and unclear lines of responsibility (a common fault is a lack of upward reporting procedures);
- Organizational evolutionary drift in which changes in organizational structure are made without reference to the original roles of people prior to change;

- Changes in business objectives or development phases. (An example is that a development project may be environmentally well handled but as operations begin total personnel change may occur over a prolonged period. Original environmental concerns and experience in handling them are lost in the drive for commissioning;
- Inadequacy of local environmental policy, i.e. head office policy is handed down without translation to meet the local needs of the plant, location, etc.

Sometimes line management have insufficient knowledge of the system operating below them. It is also common for management downwards to be clearly defined but with no feedback loops enabling grass roots environmental problems to be brought back upwards to appropriate management levels.

This can lead to a serious block in communications. In one audit in Europe we found that upward reporting was via a small local subsidiary service company whose existence was not known to any of the head office in line management in London. Consequently years of reporting of environmental problems vanished into the files of the subsidiary and had never highlighted to the only part of management empowered to take action.

Our experience with corporate audits has been very positively good. Initially there was resentment and suspicion; however with pre-audit diplomacy, incorporating business nominees in the audit team, conducting the whole operation in a spirit of cooperation and with full consultation during report writing, the audited felt that the overall result was helpful.

Audit reports are confidential to the business audited and are provided to the BP Director HSE with this understanding. In the post audit follow up period we have found that those audited were fully willing to make appropriate organizational changes in line with recommendations. In many cases such changes were made as soon as weaknesses were identified even in advance of the audit report being submitted.

It was also common to find that in the soul-searching that is inevitable prior to the arrival of the audit, problems were identified and rectified. Often all that is needed is the focus of attention.

Most audit reports are quite detailed and record the conclusions from head office interviews and interviews done on sample locations. As a general rule we find that businesses and activities that are environmentally well run are well managed and cost effective in all their operations. So much so that environmental and safety audits are a good barometer of management standards generally. This experience is shared by other major companies including some which are US-based.

7. BUSINESS INTERNAL ENVIRONMENTAL AUDITS

In recent audits the recommendations have encouraged the BP businesses to instigate their own environmental audit programmes. These are now generally in place. They have a different purpose than the corporate audit. Internal audits are intended to enable facility management to obtain an objective view of the facilities' overall environmental performance, including guidance on improvements. They are generally more technically orientated.

The internal audit reports are restricted to internal use by the facility management and their own line management. The initiation of such audits is generally the responsibility of the central business, associate, or divisional management.

There are several performance elements in the Business Internal Audit (see Appendix I), each of which is assessed in the evaluation of overall performance. These include:
- Policy: understanding and implementation;
- Regulations and compliance;
- Plant design and Operation;
- Operating Procedures and Practices;
- Maintenance Practices;

- Source Monitoring;
- Receiving Environment Monitoring;
- Oil and Chemical Spill and other Emergency Contingency Measures;
- Incident Reporting and Remedy;
- Environmental Training and Awareness;
- Contractor Selection and Performance;
- External Communication.

8. COMPLIANCE AUDITS

Compliance Audits are relatively simple to conduct although they may be time consuming.

9. SITE AUDITS

These are spot checks of sites known to have potential environmental problems. These are usually but not always remote from their business head office and typically will be located in known sensitive locations. Such spot checks cannot really be dignified with the audit title but are valuable for a rapid check on environmental conditions.

10. ASSOCIATE AUDITING PROGRAMMES

In all areas in which BP is active there is a National Associate Company, e.g. SFBP, BP Brazil, BP America. These Associate companies act as the agent for the businesses in the various countries in which the group has interests or operations.

It is the associates' role to look after the representation of the BP Group in Government and Public Affairs. In many respects the associate acts as corporate office and it will have its own policies and guidelines for the business operations. The role of the National Associate is an essential component of the matrix management by which the BP Multinational Group is able to develop its business interests with appropriate National Identities within a highly complex international framework. Within any corporate or business environmental audit the links with and role of the national associates will be examined. By the same token each National Associate has its own clear responsibilities to the Corporate Head Office, to the country in which it operates and to the businesses for which it is the national agent. There will be an associate environmental policy. The associates therefore also conduct environmental audits which are in effect similar to those of the Corporate Head Office. It is a Group aim to encourage National Associates to initiate monitoring programmes. As this develops the need for standard corporate audits will diminish and greater attention can be given to issues, etc.

Corporate Audits of the associates themselves may be needed from time to time.

The matrix management of multinational companies makes any national audit legislation difficult to apply and conduct. The need for an established, reliable internal audit matrix to match that of the management matrix is obvious. Such audits are evidence of the commitment which is ensured in Group policy with its implementation principles.

11. METHODS USED

Guidelines
The BP environmental auditing guidelines comprise eleven performance areas, divided up between management and technical issues. Under each of these eleven areas there are a number of performance criteria. These are generally worked in a way which calls for professional judgement rather than a yes/no approach. The guidelines are hence not a substitute for an experienced audit team.

Auditing is carried out by interview, site inspection and examination of documents. All management performance areas should be covered comprehensively, but because of time constraints it will often not be possible to look comprehensively at all technical areas. The basic approach should hence be:

1) Ensure that all the management performance areas are fully assessed in interviews and relevant documentation is examined.
2) Assess all the technical performance

areas by interview, site inspection and examination of documentation, but accept that in some cases performance criteria will not be examined in depth. Normally the technical issues which should receive priority attention will be decided before the audit begins but some flexibility in this respect is necessary. It is essential that all technical performance areas are assessed to some extent during the audit.

The information gathered on technical areas may not be comprehensive in all cases, but, in general, will serve as a test of the management systems assessed in 1) above. The audit system hence aims to ensure that management procedures and attitudes are correct and to test, as far as possible, that this is reflected in the technical performance of the installation.

12. ON-SITE PROCEDURES

i) Interviews
Interviews with management normally concentrate on the management performance areas (1-4) and those with supervisors and operational staff on the technical areas (5-11). In many cases the latter interviews can be carried out during site inspections.

ii) Site inspections
The areas to be inspected will frequently by obviously "critical" environmental ones, e.g. effluent treatment areas, land restoration areas, etc. It is often valuable, however, to inspect other areas to see how matters are handled away from the immediate focus of environmental attention. Candidates here could include fire training areas, contractors' lay-down areas, temporary office/accommodation areas, etc.

The timing of site inspection will frequently be a matter of convenience but, in general, it is preferable to visit areas when relevant operations are in progress, e.g. visit a pig-trap when a pig is being removed. It is also valuable in some cases to visit areas before they are discussed with supervisory staff so that specific problems observed can be raised in discussion. Alternatively, when necessary, staff should be re-interviewed following site visits.

iii) Inspection of documents

Certain documentation should always be inspected thoroughly, in particular:

- all regulations/permits;
- monitoring data;
- emergency plans;
- policy documents.

It will not be possible to thoroughly inspect all other documentation, so it will be necessary to prioritize and sample. Samples of the following types of document should normally be examined:

- waste records;
- operating procedures;
- standing instructions;
- HSE committee minutes;
- management committee minutes/ agenda;
- maintenance procedures;
- site publications, newpapers etc. ;
- operating plans;
- incident records.

iv) Audit Team

An audit team will include the experience and expertise necessary to implement the guidelines and should have credibility with facility management. It should be able to recognize the balance between conducting a rigorous audit and not unnecessarily interfering with facility operations.

We have found that the team should be large enough to allow exchange of ideas and development of consensus views. An audit cannot properly be carried out by a single individual. The ideal team size is three to four people,

depending on the size of the facility, and should include two people with expertise and experience in environmental matters and one or more who is familiar with the type of operation being audited. The audit team should not include a member of staff of the facility being audited, although the facility should be asked to nominate a member of staff to assist in making the arrangements for the audit, i.e. arranging interviews, site inspections, etc.

13. CONCLUSIONS

Environmental audits are a fully integrated component of Environmental Protection Management in the BP Group. They are voluntarily conducted in response to a need to ensure that company policy has some strength and is seen to be important.

The experience to date shows that audit report recommendations have been speedily complied with and have resulted in improved management awareness and performance.

Environmental audit approaches must be flexible and tailored to meet the nature of the business or unit being audited. They tend to be evolutionary in nature, adapting to new business situations and the inevitable changes that occur in the external and internal business environment. It is not possible to produce meaningful mandatory audit approaches, although guidelines have been helpful (see Appendix). Audit approaches appropriate for the BP structure may be inappropriate in other organizational frameworks.

CHAPTER 4 - APPENDIX 1:
AUDITING PERFORMANCE ELEMENTS

1. POLICY

Managers and supervisors should be aware of, understand and implement the business Environmental Policy Statement. Individual facility policies setting out the local arrangements for compliance with the company policy should be available, and there should be evidence that facility employees are aware of and understand the company and facility policies as they relate to their particular duties.

2. REGULATIONS

Appropriate facility personnel should know and comply with the national and local environmental regulations and Company Standards which are applicable. Up-to-date copies of these should be readily available, should be understood, and the implications of the requirements should be known to facility management. All necessary environmental consents and permits are current and copies should be available.

3. PLANT DESIGN AND OPERATION

The facility plant should be designed with technology appropriate to the achievement of legislative and Company environmental standards incorporated. All specific environmental protection equipment installed should perform well such that environmental standards are met. Facility management should be aware of the latest technological advances in environmental protection appropriate for their operation. There should be a continuing commitment to review and incorporate appropriate new technologies into the management and operation of facility activities. In a cost effective manner this ensures an evolving upgrading of environmental protection which avoids crisis rectification.

4. OPERATING PROCEDURES AND PRACTICES

Written operating procedures which include emission, effluent and solid waste limitations as required by conditions, consents, regulations and standards should exist, be understood, and be followed for pollution control systems and processes. Operating procedures for other major operations should contain environmental protection instructions as required. Other operating practices should not give rise to environmental concern.

5. MAINTENANCE

Maintenance practices should provide prompt corrective action to deficiencies in equipment resulting in environmental risk. The preventative maintenance programme should include all environmental protection equipment. The potential environmental implications of maintenance activities, including shutdowns, should be considered during planning stages and proper management of wastes and effluents should be provided at such times. Maintenance standards should be high, not resulting in environmental impairment.

6. SOURCE MONITORING

The quantities and qualities of wastes and emissions from Company operations should be monitored regularly and the status of compliance with applicable laws, regulations, consent and Company standards should be demonstrated by these monitoring programmes. There should be means of ensuring that appropriate action is taken in the event that monitoring reveals that waste streams or emissions are not in full compliance with consent limits or standards. Monitoring techniques and results should be verified by routine quality assurance checks and organized records should be kept of the analyses and equipment or method calibration.

7. RECEIVING ENVIRONMENT MONITORING

An effective programme of monitoring of the effects of wastes and emissions from Company operations upon the receiving environment should exist. The monitoring programme should reflect any statutory requirements and the recommendations of the Environmental Impact Assessment of the facility in terms of scope and sampling frequency. If monitoring has indicated environmental perturbation in excess of that predicted by the Environmental Impact Assessment or in excess of that permitted by legislative or Company standards, appropriate investigative and remedial action should have been taken.

8. OIL AND CHEMICAL SPILL AND OTHER EMERGENCY CONTINGENCY MEASURES

Written, up-to-date response procedures including contingency plans and resources covering incidental discharges should be available, accurate, and able to be implemented in a timely and competent manner. Appropriate training of personnel and exercising response plans should be undertaken.

9. INCIDENT REPORTING AND REMEDY

Records should be available detailing all incidental discharges. Reports should also be prepared and available for other environmental incidents e.g. smoky flares, nuisance noise, odours, etc. Each report should include the status of any internal or external follow-up activity. Effective steps should be taken to study and remedy the fault. Written investigation reports should be prepared and indicate the corrective action taken

to prevent recurrence.

10. ENVIRONMENTAL TRAINING AND AWARENESS

Appropriate environmental instruction should be included in personnel training programmes. Facility staff should be aware of the environmental implications of operations in their area of responsibility. Adequate staff resources should be available to the facility in the fulfilment of Company environmental policy and statutory requirements. Line management should be active in promoting/disseminating environmental information. There should be effective flow of environmental information between environmental staff and the facility and within the facility. Appropriate employees should have access to environmental technical literature and publications, and participate in professional environmental organizations and continuing educational programmes.

11. CONTRACTOR SELECTION AND PERFORMANCE

Outside contractors should perform in an environmentally sound manner under knowledgeable supervision since their activities reflect upon the Company and may be its ultimate responsibility. Reputable contractors should be chosen having due regard to their environmental record. Contractors should be made aware of BP's environmental policy; contracts should clearly state the contractor's environmental responsibilities.

Periodic monitoring of the environmental aspects of contractor work and activities should take place, with results fed back to the contractor selection procedure.

12. EXTERNAL COMMUNICATION

There should be evidence of clearly defined routes and responsibilities for liaising with the local community and/or interest groups and developing an acceptable facility image based on proven environmental performance. There should be a system for responding effectively to complaints and any reasonable requests for information. The facility may participate where appropriate in projects promoted or organized by local and national environmental bodies or by the Corporate Head Office of the Group.

In addition to the forms of audit discussed above there are other environmental audits which are conducted. These include Regulatory Compliance audits in which an investigating team checks the status of compliance with regulations, guidelines and legislation.

CHAPTER 4 - APPENDIX 2: GUIDELINES

1. Policy, Responsibilities and Organization (Management)

Performance Criteria	Auditors Notes
1.1 There is a written document endorsed by senior management, detailing the local arrangements for implementation of the environmental aspects of BP Group HSE policy.	Verify policy. How distributed? How current? Available in Business/ Associate HQ?
1.2 All staff are familiar with and accept their responsibilities for environmental matters.	Check at all staff levels.
1.3 A member of management is formally designated as the leader or coodinator of environmental protection programmes.	Job description.
1.4 Sufficient manpower resources are available for implementation of environmental protection programmes.	
1.5 Management structures are such that an adequate flow of information on environmental matters reaches senior management.	Management awareness of main issues. Reporting structure.
1.6 Adequate functional links exist with other parts of the BP Group on environmental matters, e.g. Business HQ, National Associate offices or Group supporting services.	Documentation received at site, e.g. "green book". Meetings held. Familiarity HQ staff with site issues. Availability in HQ of contingency plans regular reports etc.

2. Procedures (Management)

Performance Criteria	Auditors Notes
2.1 Environmental issues are included in long and short range planning, including:	
i) Establishment of written objectives/targets and assessment of performance against those targets.	Examine objectives. Staff appraisals.
ii) Longer term and overall strategies for environmental protection, e.g. energy conservation programmes, waste minimization programmes and land management programme.	Examine documentation.

2.2 Long and short term planning take account of recommendations made during environmental assessment of the project.

Familiarity with environmental assessment.
Operating plans.

2.3 There are procedures for evaluation of the environmental impact of:
i) New facilities and processes.
ii) Modifications to facilities or processes.

Capital Release procedures.
Consultation with HQ etc.
Lower level of capital at which procedures work.

2.4 There are procedures to ensure environmental safeguards in the event of modification removal, demolition or decommissioning of facilities.

Involvement Environmental Co-ordinator.
Ground contamination.
Wastes arising.
History of waste disposal practices.
Remedial Actions.
Finance.
Revised plans/operating procedures.

2.5 There is an internal review programme designed to identify and evaluate environmental hazards at regular intervals.

Look for planned programmes.

2.6 There are procedures to ensure that environmental problems are not caused by contractors, including:

Examination contractor policies/records.

i) Screening prior to selection.

ii) Contractors made aware of requirements regarding environmental performance.

BP HSE policy to contractor other instructions/guidance.

iii) Monitoring of environmental performance.

Environmental awareness of BP supervisors.

iv) Appropriate contract conditions.

Limitation of liability.

v) Reporting from contractors on environmental issues.

Written or informal.

2.7 Environmental issues are included in regular reporting to Business HQ, Associate H.O. etc. and an annual report on environmental matters is prepared.

Check documentation.

2.8 Records and reports of incidents, monitoring data, waste disposal documentation and other environmental information are maintained in a well organized and accessible form, and are available to the environmental co-ordinator.

Monitoring records.
Communications with regulators.
Waste disposal documentation.
Incident reports and investigations.
Regulations.

2.9 There are procedures to ensure that:

i) Written operating instructions for major operations contain the necessary up-to-date environmental protection instructions.

Examine operating instructions:
– Environmental Protection Plant.
– Other plant.

ii) Regulatory constraints are incorporated in operating instructions.

Information flow from Environmental Co-ordinator to operations.

iii) Where there are no regulatory restraints, appropriate BP standards are incorporated in operating procedures.

Out-of-compliance discharges.
Spillages.
Odours.

2.10 There are procedures in place for investigating and taking remedial action in the event of environmental problems occurring.

3. Training and Awareness (Management)

Performance Criteria

Auditors Notes

3.1 There is an effective routine flow of environmental information among all appropriate levels within the facility, including:

i) Through formal mechanisms, i.e. line management and committee structures.

HSE committee minutes.
Management committee agenda/minutes.

ii) Through inclusion in site publications, technical meetings and other general mechanisms for information flow.

Site newpapers.
Technical meeting agendas.

3.2 Management, supervisors and staff are aware of the environmental issues which are relevant to their areas of responsibility.

Assess level of awareness in relation to responsibilities.

3.3 All staff have an appropriate level of awareness of regulatory requirements.

3.4 Employees responsible for managing and implementing environmental protection programmes are trained and qualified to the appropriate level.

Training of:
    E. Coordinator
    Management
    Supervisors
    Operating Staff.

3.5 Programs are implemented which give an appropriate background of environmental awareness for all employees.

Induction courses.
Inclusion E. in other training courses.

3.7 Where necessary site guidelines detailing particular environmental sensitivities, regulations etc. are available to all staff.

Site videos
      booklets.

3.8 Key staff are aware of probable regulatory and technical developments in the environmental field.

3.9 The site is represented on local regional or national associations where environmental issues are discussed.

Activity in working groups etc.

4. External Relations (Management)

Performance Criteria	Auditors Notes
4.1 There is an effective and clearly defined system for liaising with regulatory authorities at the local, regional and national levels.	Points of contact.
4.2 There is an effective and clearly defined system for liaising with the Press and other media, local community and interest groups and the general public, using the services of public affairs specialists where necessary.	Press releases. Interviews. Routine and emergency situations.
4.3 Local communities and interest groups are adequately informed of the environmental impact of the installation, using environmental monitoring data when appropriate.	
4.4 Complaints are investigated and recorded in a systematic fashion and appropriate actions taken.	Complaints records. Feedback to complainant.
4.5 Where necessary, there are systems for alerting the public of imminent hazards.	Air pollution alerts. Drinking water alerts. Spillage alerts.
4.6 There is an appropriate level of participation in projects promoted by local or national bodies.	

5. Regulations and Monitoring (Technical)

Performance Criteria	Auditors Notes
5.1 All necessary permits are up-to-date, reflect actual operations and copies are available.	Check documents.
5.2 Where there are no regulatory requirements, appropriate written standards have been developed internally.	
5.3 All emissions monitoring required by regulatory authorities is carried out using satisfactory procedures.	Air Water Waste Sewage – sampling – analysis
5.4 Where there are no regulatory requirements for emissions monitoring, there is a means of regular feedback on the efficiency of environmental protection measures.	

5.5 Appropriate receiving environment monitoring is carried out according to statutory requirements or other defined objectives and uses satisfactory procedures.

Methods
    Sampling
    Analysis
    Calculation
    Etc.
Community Noise

5.6 Where applicable, the objectives of receiving environment monitoring are linked to the recommendations of project environmental assessments or other environmental quality objectives.

Check Environmental Assessment

5.7 Records of all monitoring data are available and have been reported to regulatory authorities as required.

Check documentation
– emissions monitoring
– receiving environment monitoring

5.8 Steps are taken to identify the causes and take remedial action when:

Check documentation

i) Monitoring data or other information indicate non-compliance.

ii) Receiving environment monitoring indicates unacceptable environmental degradation.

6. Emergency Planning and Response (Technical)

Performance Criteria

Auditors Notes

6.1 There are emergency plans covering all the significant risks on-site and available to all relevant personnel.

Ensure coverage of all emergency sources.
Distribution of plan.

6.2 The emergency plans:

i) Clearly define responsibilities.

Check documentation.

ii) Clearly define all procedures.

Check documentation.

iii) Define environmental sensitivities.

e.g. sensitivity mapping.

iv) Define linkages to local authorities and other bodies.

Emergency Services.
Water authorities.
Conservation bodies.

v) Define linkages to other plans or resources.

Other BP sites.
BP HQ plans/resources.
Wider industry resources.

vi) Are regularly updated.

Updating procedure.

6.3 Key personnel are familiar with the provisions of the plan.

Ensure staff understand own role in plan.

6.4 Equipment and manpower of adequate quantity and quality are available for emergency response:

Inspect equipment
compare with plan

- from on-site resources

- from identified local, national and BP resources

6.5 Response equipment is well-maintained and accessible.

Maintenance programmes.
Inspect equipment.

6.6 There is an established programme for the training of personnel in the use and handling of emergency response equipment.

6.7 There is an established programme of emergency plan exercises, with systems for implementing resultant recommendations.

Examine reports of exercises.
Check in annual objectives.

6.8 Incident review procedures are formalized and recommendations are implemented.

Check incident reports.

6.9 All incidents are recorded and reported, as appropriate, to management and regulatory authorities.

Check records.

7. Pollution Sources and Minimization – Air and Water (Technical)

Performance Criteria

Auditors notes

7.1 All sources of air pollution are identified and quantification undertaken where necessary.

Fuel Quality.
Emission Inventories.
Verify in site inspection
– process areas
– storage areas
– stacks heights/positions vents.

7.2 Plant is well designed, inspected and maintained so as to minimize fugitive emissions to atmosphere and leaks of liquids.

Gain impression in site inspection.
Include noise.

7.3 Plant is operated so as to minimize emissions to air and water.

7.4 All sources of surface water pollution are identified and quantification undertaken where necessary. All relevant drainage plans are available.

Drainage plans.
Fire water.
Sewage.
Stormwater.
Cooling water.
Verify in site inspection.

7.5 Drainage facilities are adequate and well maintained.

Verify in site inspection.
Drain destination known.
Environmental sensitivities understood.
Opportunities for containment.

7.6 Systems for containment or prevention of spillage and leakage are adequate.	Tank level controls/alarms. All necessary areas bunded – main storage areas – drum storage areas – waste storage areas – process areas – vehicle loading/unloading areas Lagoons/ponds – lining – level control – alarms
7.7 Appropriate systems for prevention and monitoring of leakage are installed and effective.	Integrity metering. Corrosion prevention.
7.8 Off-site measures are taken, where appropriate, to minimize the probability of on-site pollution events.	Inspections of vehicles, railcars. Tanker check-lists.
7.9 Environmental data is available about all substances which could enter waste streams.	Examine documents. Chemicals used on-site. Chemicals entering site with raw materials.

8. Pollution Treatment and Discharge – Air and Water (Technical)

Performance criteria Air	Auditors notes
8.1 Treatment facilities are well designed to achieve the required standards and discharge locations are well sited – Air – Water	Check during site inspections.
8.2 The operation of the treatment facilities results in emissions within regulations of other standards – Air – Water	Monitoring data.
8.3 Preventative maintenance is carried out on treatment facilities.	Discuss with maintenance staff. Check in site inspections.
8.4 Written operating procedures for the treatment facilities indicate the action to be taken if regulatory or other standards are exceeded.	Check operating procedures.
8.5 There is satisfactory provision for treatment and discharge of liquid and atmospheric wastes during maintenance or other shutdowns. – Air – Water	

9. Pollution Sources and Minimization – Waste and Groundwater
(Technical)

Performance Criteria	Auditors Notes
9.1 Environmentally hazardous materials are stored on site in a manner commensurate with the hazards they pose and with regulatory requirements.	Check on-site. Regulations.
9.2 Appropriate environmental data are available concerning hazardous materials on-site.	Check records.
9.3 Wastes are properly identified, cagegorised and separated.	Check on-site. All sources considered – from operations – from maintenance/overhaul – from construction – from decommissioning/demolition.
9.4 Wastes are stored in a manner commensurate with the degree of environmental hazard they pose and with regulatory requirements.	Waste storage arrangements.
9.5 There is a programme for detection and control of losses from underground storage tanks.	Corrosion prevention. Integrity testing. Soil groundwater sampling. Check on-site.
9.6 Systems exist to detect leakage, from drains, lagoons, landfills, land farming areas, tailings ponds or similar potential sources of groundwater pollution.	Soil/groundwater sampling. Check on-site.
9.7 There is information available about the status of the site in respect of groundwater contamination.	Monitoring data. Boreholes.
9.8 The sources and extent of any groundwater contamination have been identified and appropriate remedial actions are being taken.	

10. Pollution, Treatment and Discharge – Waste and Groundwater
(Technical)

Performance Criteria	Auditors Notes
10.1 Appropriate methods of waste treatment and disposal are in use for all types of waste generated on-site.	Verify methods used.
10.2 Waste treatment, transport and disposal activities are documented adequately to: i) satisfy regulatory requirements;	Check documentation.

ii) allow appropriate tracking of waste consignments;

iii) give protection from liability by forming a historical record of on- and off-site disposal activities.

10.3 The activities of contractors engaged in waste disposal are regularly checked by visits to landfill sites or other waste disposal/treatment facilities.	Check records.
10.4 Contract conditions specify the type of treatment/disposal to be used.	Check records.
10.5 Waste treatment and disposal or storage facilities on-site are well designed and operated to meet regulatory or other standards.	Landfills. Incinerators. Temporary storage areas.

11. Land Management (Technical)

Performance Criteria	Auditors Notes
11.1 Restoration of disturbed areas is carried out according to a plan which is acceptable to local communities and regulators.	Check plans and discussions with authorities.
11.2 Areas to be used for storage and restoration of inert waste are clearly identified.	Check plans.
11.3 Restoration is carried out to a high standard:	Visit revegetated/profiled areas
– grading/profiling	– tank farms
– soil treatment/seeding	– road verges
– soil erosion	– spoil heaps
11.4 Efforts are made to overcome difficulties in restoring difficult substrates, e.g. mine spoil, tailings.	
11.5 All areas of vegetation are adequately maintained.	
11.6 The visual impact of the facility is acceptable to local communities.	
11.7 Housekeeping procedures result in a neat and tidy appearance on-site.	Check on-site.

CHAPTER 5 - ENVIRONMENTAL AUDITING, UNOCAL CORPORATION

Michael L. KINWORTHY
Unocal Corporation

BACKGROUND
PROGRAM PURPOSE

The objectives of Unocal's Environmental and Health Compliance Review Program are to provide management and the Board of Directors with verification that:

- the company's operations are in compliance with applicable governmental and internal health and environmental requirements;
- systems are in place to assure that compliance continues; and
- operating facilities and process units are designed, constructed, operated and maintained to protect employees, communities and the environment.

To achieve these objectives, Unocal's program employs both detailed data-gathering and on-site audits. Each operating facility completes an annual compliance questionnaire that is prepared especially for it. Pertinent questions are asked dealing with air, water, waste, toxics, etc. to identify environmental compliance. Data are gathered on over 2300 operating facilities and provide assurance of compliance to Unocal's Board of Directors before it submits its 10K report to the U.S. Securities and Exchange Commission.

This data-gathering program is supplemented with on-site audits conducted by select teams at specific sites each month. Compliance audits are comprehensive examinations of the programs, procedures and practices of operating locations. An audit verifies the degree of compliance with governmental and internal requirements. An evaluation of the design and implementation of the location's compliance management system is included in each audit.

ORGANIZATION AND STAFFING

At Unocal we are organized so that the Environmental and Health Compliance Program ultimately reports to the Vice President, Health, Environment and Safety. Responsibility for overseeing the implementation of the program rests with the Manager, Environmental Programs and with the Company's Environmental and Health Compliance Program Management Committee. That committee is comprised of representatives from the following divisions, subsidiaries and departments:

- Unocal Chemicals Division;
- Unocal Energy Mining Division;
- Unocal Geothermal Division;
- Unocal International Oil and Gas Division;
- Unocal Oil and Gas Division;
- Unocal Sciences and Technology Division;
- Unocal Refining and Marketing Division;
- Molycorp, Inc.;
- Environmental Sciences Department;
- Industrial Relations Department;
- Law Department;
- Medical Department.

The Management Committee elected to implement the program through the use of in-house review teams. Comprised of Company personnel, the teams conduct on-site compliance reviews of Unocal facilities and operate under the direction of a team leader appointed by the Manager of Environmental Compliance.

The team members have varied functional backgrounds including occupational medicine, industrial hygiene, environmental, research and process engineering. Each member is assigned a discipline (i.e. air quality) to be responsible for during the review.

The Management Committee actively supports the audit program by: (a) providing guidance on the interpretation and applicability of governmental and internal requirements; (b) proposing revisions to audit procedures and protocols; (c) participating in auditor training sessions; (d) resolving differences of opinion raised by the audited locations and (e) reviewing the adequacy of action plans submitted by the facilities in response to final audit reports.

COMPLIANCE AUDIT PROGRAM SCOPE

All world-wide operating locations of Unocal are included within the scope of the compliance audit program. The performance of each location is audited against:

- governmental requirements (national, state, and local); and
- internal requirements (corporate, divisional, and facility policies, procedures and standards).

The functional scope of compliance audits includes environmental and health issues. Each issue contains a number of more specific functional areas. Here are some examples:

- Health:
  · Hazard Communication
  · Industrial Hygiene
  · Occupational Medicine
  · Injury, Illness and Accident Recordkeeping and Reporting
- Environmental:
  · Air Pollution Control
  · Water Pollution Control
  · Groundwater Pollution Control
  · Spill Prevention
  · Solid and Hazardous Waste Management

· PCB Management
· Toxics Control
· Hazardous Waste Site (Superfund) Management

AUDIT SCHEDULE

The number and diversity of Unocal operations requires an orderly approach to scheduling. The primary factors include assessments of the risks associated with potential incidents, such as the release of toxic materials, and the potential compliance risks associated with departures from applicable requirements.

Risk classifications have been completed for all operating locations. Those assessed to have higher risk potential are scheduled for more frequent audits.

When developing the audit schedule to obtain a sample of the lower risk locations, consideration is given to:

- potential environmental and health issues;
- facility size;
- past environmental and health performance; and
- selection of representative locations from all business and geographical areas.

This balanced approach, focused on locations of higher risk potential, while maintaining an overview of all environmental and health activities of the Corporation, allows the utilization of audit resources to be effectively prioritized.

AUDIT METHODOLOGY

A five-step approach is utilized in the conduct of compliance audits and involves:

- developing an understanding of the location's compliance management systems;
- assessing the strengths and weaknesses of the compliance management systems;
- gathering audit evidence;
- evaluating audit evidence; and
- reporting findings.

When followed in accordance with established procedures, this approach allows a team to independently obtain the information required to evaluate the design and effectiveness of a location's management system and to verify compliance with governmental and internal requirements. The audit findings form the basis for a judgement of the location's performance relative to the scope of the audit.

The audit steps are summarized as follows:

Step 1: Understanding Compliance Management Systems

The first step the audit team takes is to understand a location's compliance management systems. This provides the audit team with a framework for evaluating the effectiveness of the location's systems and compliance performance in subsequent audit steps.

Understanding the location's compliance management systems begins early. We review policy and procedure manuals and organization charts, we study descriptions of the location and its operations, and we learn the responsibilities of key personnel.

The process continues during the on-site audit. The actual audit begins with a meeting of the audit team, the location manager and appropriate staff people at the location. The team leader opens with an overview of the Environmental and Health Compliance Program and discusses the purpose and scope of the planned audit. The facility manager then briefly describes the general operations, programs, resources, management systems, and significant issues that relate to the audit.

The audit team and key staff then tour the location. This orientation provides the audit team with a "hands-on" introduction to the size, layout, operations and facilities of the location.

After the orientation tour, the audit team proceeds with an internal controls questionnaire to obtain even more specific information about the management systems. This questionnaire also helps determine whether certain requirements apply to the location's activities.

The combination of the information obtained before the audit, during the opening meeting and location tour and from the internal controls questionnaire, offers the auditors more detailed understanding of the compliance management system assosciated with each assigned functional area. In-depth interviews with key people, additional tours of specific sites and a critical review of practices and procedures augment this understanding.

This step of the audit process results in a thorough documented understanding of a location's overall management system and its functional area management systems.

Step 2: Assessing Compliance Management Systems

The second step in the audit process is to assess the strengths and weaknesses of the location's compliance systems. It is also the time to develop appropriate strategies for verifying compliance.

The objective of the assessment is to determine if the location's internal management systems are:
- adequate and appropriate to achieve and maintain full compliance; and
- capable of protecting the Corporation's interests while considering the nature of the risks associated with the location's activities.

This assessment provides the auditors with a context for understanding health and environment exceptions in relation to potential weaknesses in the management systems used by the location to govern or direct health and environmental activities. The audit team can then communicate the audit findings to management more effectively. They can also offer observations on the underlying causes for health and environmental exceptions noted during the audit.

Step 3: Gathering Audit Evidence

The evidence gathered in this step verifies the functioning of location management systems and allows assessment of compliance with governmental and internal requirements. Audit evidence is gathered in accordance with program objectives; that is, to substantiate compliance as well as to identify exceptions. The audit team utilizes written audit protocols as guides. These protocols relate specifically to programs or practices within the scope of the audit and generally describe how each subject is to be examined.

In gathering audit evidence, the auditors will generally (a) review records, reports, files and other documentation maintained by the location; (b) conduct interviews with facility staff; and (c) physically inspect facility operations, equipment, and activities related to the scope of the audit.

An attempt is made to obtain corroborating evidence from independent sources whenever possible. Because of the large number of application requirements and the amount of information related to compliance performance, evidence gathering is generally a sampling activity rather than a thorough review. Based on the evidence obtained, location practices are compared with governmental or internal requirements.

At the conclusion of this step, the protocols are reviewed to ensure that: (a) all appropriate auditing steps were complete, (b) the testing plans were appropriate, and (c) the evidence gathered was sufficient to substantiate conclusions as to compliance or any exceptions noted.

Step 4: Evaluating Audit Findings

In this step of the audit process, the evidence gathered is reviewed and evaluated. A complete list of all exceptions is prepared and this list is analyzed to identify any trends or patterns. The number and severity of the exceptions are then evaluated. The audit team then offers a judgement on compliance performance relative to the scope of the audit. At this time, the audit team also prepares a list of observations (i.e., areas of concern related to practices not subject to either governmental or internal requirements) noted during the audit.

Step 5: Reporting Audit Finding

The final step of the audit process is to formally communicate results to the audited location and appropriate levels of Unocal management.

This communication begins at the exit meeting in which the audit team first discusses all findings with local management. This meeting ensures that all findings are understood. It also resolves, it possible, any outstanding issues and potential misunderstandings.

Following the on-site audit, the team prepares a written report which clearly comunicates audit results to appropriate levels of Unocal management.

The team's draft report is sent to the facility management which has forty (40) days in which to prepare and submit a response to the Manager, Environmental Programs. The report is then finalized and released to that operating division's senior management.

The line organization is responsible for developing and implementing corrective action plans in response to all audit findings. All action plans are reviewed by the Manager, Environmental Programs. He tracks implementation of corrective actions through periodic status reports submitted by the line organization. Actual completion of action plan items are verified during subsequent audits.

ENSURING AUDIT QUALITY

A number of quality control procedures ensure audit quality.

First, knowledgeable people conduct all compliance audits. They have been thoroughly prepared for their assignment.

Second, all Unocal compliance audits follow written protocols. The protocols reflect the objectives of the audit. They methodically guide the team through the desired examination. They contain specific audit tests and procedures and outline desired audit documentation. In addition, they contain questionnaires and checklists to guide the audit team in conducting certain protective steps.

Third, all audits undergo a series of internal control checks. These are initiated during the on-site audit process with informal team reviews. At the completion of the audit, the team leader reviews the protocols and working papers to confirm that all audit steps were completed, adequate evidence substantiates the findings and established audit program procedures were followed.

Finally, an in-house attorney takes part in the review of all reports before their release to the location management.

CONCLUSION

Unocal management finds a number of benefits result from the implementation of the Environmental and Health Compliance Program. These include: assurance to Senior Management and the Board of Directors that facilities are in compliance or on compliance schedules; improved overall environmental and health performance; increased employee awareness and acceptance of their environmental and health responsibilities, and improved company public image.

CHAPTER 6 - A SYSTEMATIC APPROACH TO ENVIRONMENTAL AUDITING IN NORSK HYDRO

Per A. SYRRIST
Norsk Hydro A.S.

1. ENVIRONMENTAL AUDITING

The Norsk Hydro Environmental Audit is to be carried out in order to give the Corporate Management an overview of the environmental status of each of the operating units of the Hydro group of companies. The Corporate Health, Environment and Safety Section (HES) is responsible for organizing such audits.

The objectives of the audit are to verify, through study of available documentation, interviews and spot checks:

1) that Norsk Hydro Corporate Management environmental objectives and requirements are known throughout the Company;

2) that actions are taken to meet these objectives and requirements; and

3) that the objectives and requirements are successfully achieved.

This implies assessment of compliance with Norsk Hydro environmental objectives and requirements in (a) Division headquarters, (b) National Company headquarters and (c) at the site, concerning import, production, storage, handling and export of materials and products.

At the site in question the audit should also verify:

4) that an environmental management system is implemented; and

5) that this system is functioning as intended.

In particular, the audit at the site should focus on the following:

– how environmental goals are established, what they are, and how they are communicated from the head office through the line organization at the site;

– how responsibilities are delegated and whether these responsibilities are delegated consistently through the organization;

– how work tasks for improved environmental management are identified and scheduled and the contents of the current environmental improvement programme;

– the availability of written procedures, and whether these are known and followed by those concerned;

– what environmental monitoring practices apply at the site and whether these are working satisfactorily;

– the environmental awareness and attitudes of management and employees.

The audit should be carried out in such a manner that it assists local management in identifying weak elements and possible improvement actions. The audit findings should be further considered by the operating unit for inclusion in the environmental action plan.

The audit will be concerned with auditing or organizational and administrative matters as well as performance of the process and effluent treatment plants.

2. AUDIT TEAM

Normally the audit team should consist of:

1) A production manager or a site manager from a similar site (as team leader);
2) A representative from HES (as experienced system auditor and team secretary);
3) An environmental and safety manager from a similar site;
4) An engineer having special knowledge of the operations under consideration.

The environmental and safety manager at the site being audited should act as site coordinator and assistant to the audit team, and may join the audit team with observation status.

The President or others from Corporate/Division Management may take part in the audit/parts of the audit, as additional audit team member(s).

3. AUDIT

Every major site within Norsk Hydro should be audited once every two years.

The audit interviews and spot checks at site should, for a typical site, be completed within three to five working days.

4. AUDIT PROGRAMME

4.1. Scheduling

The audit comprises the following elements which have to be taken into consideration in the scheduling:

1) Audit preparations:
   – timing
   – selection of audit team
   – preparatory meeting
   – collection and study of documents
2) Audit of site (three to five days):
   – briefing sessions for audit team and site management
   – interview sessions
   – area audits
   – reporting session
3) Audit of National Company Management (one-half day):

   – interview sessions focusing on strategic planning and priorities
   – reporting session
4) Combined interview and report back session with Division Management (two hours)
5) Draft and final written reports.

4.2. Preparation for auditing

HES is responsible for preparing for the audit, i.e.:

– setting dates for the audit (in co-operation with the parties being audited),
– selecting the audit team,
– requesting information from the site and distributing this to the audit team in advance of the audit.

A preparatory meeting with the site management should take place about a month in advance to discuss:

– Norsk Hydro and site goals and objectives,
– scope of the audit,
– agenda for the audit, who to interview, areas to focus on, who should participate in the report back session at the end, etc,
– information to other members of staff at the site in advance of the audit,
– documents for distribution to the audit team.

Preferably a high-ranking member of Division Management in Oslo should participate in this meeting together with the audit team secretary.

4.3. Introduction to the site

As an introduction to the site prior to the actual audit, the Site Manager should brief the audit team on operational and environmental matters. Preferably this briefing session should be combined with a tour of the site.

4.4. Audit interviews

The interviews with Division Management, National Company Management, Site Management and Supervisors, and with Employees should focus on questions outlined in standard checklists. Whenever feasible the interviews should be combined with a tour of the work area and a physical demonstration

Chapter 6 - A Systematic Approach To Environmental Auditing in Norsk Hydro on how the problems are being tackled.

4.5. Area audits
The area audits should consist of:

1) an interview session with area management, and
2) spot checks during a walk through the area.

The area audits should address environmental management in each area and examine a few codes or practice in depth.

4.6 Assessment and reporting
Reporting of audit findings should take place as follows:

– preliminary findings will be verbally reported by the audit team:
   (1) to the Site Management at the end of the site visit,
   (2) to the National Company Management at the end of the National Company headquarters visit, and
   (3) to the Group/Division Management during the interviewing session with the Group/Division Management.
– the draft written report will be sent by HES to Site Management for commenting,
– the final written report will be published by HES as soon as possible after receiving comments and sent to the President, Group/Division Management, National Company Management and Site Manager.

The audit observations and possible improvements should be discussed at the reporting sessions. The written report should summarise possible improvements agreed upon, with reference being made to specific audit observations.

CHAPTER 7 - CENTRALIZED AUDITING IN A DECENTRALIZED CORPORATION

Linda A. WOLLEY
ITT Corporation

OVERVIEW

Environmental Compliance auditing has gone through several incarnations at ITT Corporation, mirroring the management style of the Chief Executive Officers (CEOs) and Presidents who have been in place at the time. I will seek to explain the different types of compliance auditing programs that ITT has used, describe the strengths and weaknesses of those programs, and discuss the problems of compliance auditing that are unique to a highly decentralized company.

HISTORY OF ITT'S AUDITING PROGRAMS

A form of environmental compliance auditing began at ITT in the 1960's, and was conducted by the Quality Department, headed by Dr. Philip Crosby. Dr. Crosby became nationally known as "Mr. Quality". Although not a modern-day environmentalist, Crosby advocated cleanliness, purity, wholesomeness and safety in products and manufacturing plants.

Under Crosby, ITT had a highly centralized auditing program, which was in keeping with the highly centralized management style of ITT's then-CEO Harold Geneen. Under Geneen, ITT had a large and powerful Headquarters staff that exercised considerable control over subsidiary companies (or, in ITT parlance, "units") in the areas of government and public relations, legal, health and safety, personnel, taxes accounting, and all aspects of compliance and auditing.

The Headquarters Quality staff not only oversaw the units' operations, but had the authority to recommend and enforce changes. Since large dollar amounts were not yet associated with environmental compliance, this authority didn't have the same significance that it would have today. However, the Headquarters Quality Department authority extended to all aspects of what would be considered "quality", from plant cleanliness and fire prevention to production line design. The Department's motto was "zero defects", and every part of ITT including functions like the Legal Department had set goals to meet.

This program remained in place until the mid-1970's, when two things happened. First, the environmental movement had come into its own. The U.S. Congress passed laws to control water pollution, air pollution and waste disposal, and thus, the very nature of all corporate environmental compliance programs changed. At the same time, ITT's management changed. A new CEO envisioned a far more decentralized corporation.

Decentralization meant that the Headquarters staff had less authority and that the units had more autonomy for the operations side of their businesses. Certain staff functions (legal, executive staffing, public and goverment relations etc.) remained coordinated through Headquarters. The auditing program through this phrase changed to a self-auditing one, with far less aggressive Headquarters staff review and oversight.

A few years ago, decentralization at ITT was accomplished completely. ITT sold the operations that its name had come to signify, international telecommunications, and was organized into nine management companies: Automotive Products, Electronic Components; Financial Services; Hartford; Sheraton; Defense Technology; Communications; Fluid Handling and Rayonier. But the realization that environmental compliance cost and liabilities have a great impact on all units' operations and the bottom-line, forced the reinstitution of an authoritative centralized environmental, health and safety Headquarters function. While a detailed self-audit program is still used ITT also uses a centralized audit program. What follows is a description of that latter program.

OBJECTIVES OF A CENTRALIZED AUDIT PROGRAM

The primary purpose of ITT's centralized audit program are as follows:

(1) To determine and verify compliance with federal, state and local laws and regulations.
(2) To provide assurance to management that all operations are in compliance.
(3) To identify environmental, health and safety hazards and problems.
(4) To confirm that internal environmental, health and safety information reporting and control systems are in place and functioning.

When conbined with the self-audit program, the centralized audit yields two additional benefits:

(1) It assists facility management in identifying compliance problems.
(2) It provides a vehicle for training managers on regulations, controls and management systems.

Over time, senior management hopes that the program will do additional things:

(1) Develop a basis for optimizing resources by identifying opportunities for cost reduction; and,

(2) Demonstrate substained due diligence, and thus reduce or eliminate potential corporate liabilities.

SCOPE, STRUCTURE AND BOUNDARIES OF THE PROGRAM

In order to understand environmental auditing at ITT, something must be said about the Corporation's structure. Within each of ITT's nine management companies, are "units" (or subsidiary companies). Each unit in the ITT system has an environmental coordinator. Although the environmental coordinator reports through the management of the unit, and ultimately the overall management company, there is a great deal of interaction between those coordinators and the Headquarters EHS staff. In fact, the Headquarters EHS staff assembles all of the environmental coordinators worldwide at a biennial conference, and provides environmental updates to them on a regular and frequent basis. On a parallel but complementary track, both the Headquarters legal staff and Washington office staff are responsible for keeping unit lawyers informed of legislative and regulatory developments. Since all three departments (EHS, Legal and Washington Office) report to ITT's general counsel, there is a great deal of written communication between them, which is circulated to the units.

With considerable input from the nine ITT management companies, legal and Washington office staffs, and Headquarters auditing department, the EHS corporate staff developed an audit format. The audit reviews compliance under the following U.S. laws:

– Clean Water Act
– Clean Air Act
– Resource Conservation and Recovery Act
– Superfund
– Toxic Substances Control Act
– Safe Drinking Water Act
– Occupational Safety and Health Act
– Hazard Communication Program
– Noise Control Act

- Federal Insecticide, Fungicide and Rodenticide Act
- Coastal Zone Management Act
- Ocean Dumping Act
- Rivers and Harbours Act
- Deep Water Port Act
- Hazardous Materials Transportation Act
- Surface Mining Control and Reclamation Act
- Endangered Species Act

Obviously, not every law applies to every facility, but a standarized format is used to ensure that no area is overlooked, no matter how remote its applications might be.

The Corporate EHS notifies the ITT Management Company that an audit will be performed and assembles a customized team both familiar with the unit's operations and qualified to conduct the review. The EHS staff may or may not be a part of the team. A typical team will include personnel from the following specialities: legal, environmental management, industrial hygiene, engineering, chemistry and geology. In an area where internal expertise may be lacking, outside consultants are used.

Written review protocols, or guides, have been developed for each area within the scope of the program to methodically guide the team in conducting the audit. Facility tours, interviews, and discussions are used by the audit team to gain an understanding of the facility's compliance programs, internal management control systems, and industrial management practices. Then, through inspections, data and documentation reviews, further interviews and testing, the team evaluates facility performance in relation to program objectives. Findings are reviewed with facility management before the team leaves the site. Following the audit, a draft report is prepared by the team leader with assistance from the team and issued to facility management for review and comment. A final audit report is issued to the unit facility manager, unit management, unit legal counsel, and unit EHS coordinators.

Each audit takes approximately one to two days on-site. Prior to visiting the site, however, the audit team will review the unit's self-audit forms and responses on a pre-audit questionnaire and become generally familiar with the facility's internal controls. The schedule for audits is determined by the EHS staff, with input from unit managers and legal counsel, and announced biannually. The frequency of audits depends on the degree of hazard associated with a facility's operations or the need for structured follow-up.

Each audit team has three to five auditors and the team leader responsibility rotates. The team leader is responsible for collecting data and drafting the final report.

AUDIT APPROACH

The overall steps in ITT's audit program are described below.

1. Audit planning and preparation

Several pre-audit activities are undertaken to help ensure that audits are undertaken in an effective manner. These activities include notifying the facility manager of the planned audit, obtaining background information about the site to tailor the audit scope, and selecting and organizing the audit team. In addition, the team modifies the scope of the audit, as appropriate, to include programs that are specific to an operating company, e.g., Department of Defense EHS standards, or requirements imposed a location because of contractual obligations.

2. On-site activities

To achieve are in-depth examination of compliance programs, and internal management control systems as they relate to the scope of the audit, on-site activities follow a five-step process to reflect the objectives of the audit program. This five-step approach includes:

- Step 1: Understand internal management systems and procedures. The purpose of this step is to develop an understanding of the facility's EHS processes, performance, management control systems and level and experience of staff and resources. This is done primarily through review of background information provided by the facility, a tour of the facility, administration of a general questionnaire, and various discussions with facility staff.

– Step 2: Assess strengths and weaknesses of internal controls. Following the completion of Step 1, the team assesses the strengths and weaknesses of internal controls. This step provides the rationale for conducting subsequent audit steps and provides a benchmark for reviewing results. Internal controls that appear to be sound in design are tested in Step 3 to verify that the systems are actually in place and functioning effectively.

– Step 3: Gather audit evidence. This step forms the basis on which the audit team determines compliance with applicable laws, regulation, and corporate policies and procedures. Evidence may be gathered in a number of ways, including reviews of records, interviews with facility personnel, and verification testing of the facility's adherence to regulations, guidelines and procedures. Consistent with program objectives, Step 3 goes beyond identifying problems that may exist; evidence is also collected to determine, confirm and document compliance, and the effective functioning of internal management control systems.

– Step 4: Evaluate audit findings: Once evidence gathering is complete, the audit team evaluates its findings, and observations. Before completing the on-site review, the team jointly discusses, evaluates, and finalizes their audit findings and observations into a preliminary report.

– Step 5: Report audit findings. Throughout the audit, the team members individually and collectively share observations noted during the review with facility management. This continuous feedback is designed to minimize any misunderstandings, encourage team members to organize thoughts, and give facility personnel an opportunity to participate in the reporting process. The formal audit program reporting process begins with an exit meeting between the audit team and facility management at the end of the field work. During this meeting, the audit team presents a preliminary report and communicates the observations and findings noted during review.

3. Post-audit activities

A draft audit report is prepared by the team leader with assistance from the team following each audit. This report is addressed to the facility manager for review and comment. A final audit report is issued to the unit president, facility manager, unit environmental coordinator, and unit legal counsel. The audit report contains a listing of findings and a summary that provides management with a means for interpreting the significance of the audit results and to assist them in focusing their resources on areas where improvement is most needed. Following the issuance of the final report, the unit company president designates someone to follow-up on the report recommendations.

PROTOCOLS

Protocols are used by the team members to conduct audits. These audit protocols provide audit team members with step-by-step instructions for collecting information about a facility's EHS compliance programs and practices, and potential liabilities associated with past operational and disposal practices. These protocols ensure consistency in approach from audit to audit, and reviewer to reviewer, thus, help to ensure reliability and comparatibily of audit results.

PROBLEMS AND BENEFITS

The problems of centralized auditing in a decentralized company are summarized in one word, "turf". Since the primary responsibility for environmental management rests with line managers, there is a natural suspicion of any Headquarters oversight. The value of good communications and person-to-person contact cannot be underestimated. Fortunately, there is a great deal of communication between the Headquarters EHS and Legal Departments, and the units' environmental coordinators and legal counsels. Thus, the coordinators can turn to the Headquarters staff for information, advice and guidance.

Some of the on-going problems that arise in auditing a decentralized company are the following:

- Conflicts between "line" and "staff" plans of action, accountabilities and goals;
- Lack of manpower and difficulty in assembling audit teams;
- Diversity of operations, knowledge, and controls between units;
- Disputes about who pays for which services.

The problems that centralized auditing solves, however, outweigh the above on-going problems and almost always can be solved with good communications at all levels. The benefits of centralized auditing are as follows:

- Uniformity;
- Certainty of compliance and record–keeping;
- Clear guidance and plan for corrective action;
- No surprises — all levels of management are fully informed and working toward the same end.

CONCLUSIONS

The environmental compliance auditing program at ITT works well, probably due in large part to the strong commitment to its success by top management. In meetings with senior managers, ITT's Chairman Rand Araskog repeatedly emphasizes the need for environmental protection and encourages the nine management company presidents to develop strategic and operating plans that demonstrate on-going environmental protection efforts.

It is our view that compliance audit programs that do not enjoy the support of top management are set up to fail. We at ITT are fortunate to have a committed team.

CHAPTER 8 - ENVIRONMENTAL AUDITING – A NEW IDEA IN SWEDISH ENVIRONMENTAL WORK

Richard ALMGREN
Federation of Swedish Industries

ENVIRONMENTAL WORK IN SWEDEN

Major Environmental Issues

The Swedish Parliament established the environmental policy for the 1990s as late as June 1988. Some major problem areas are identified below.

The emissions of carbon dioxide must not in the future exceed the 1988 level — a decision that should be viewed as a contribution to avoid climate changes.

Swedish use (0.5 per cent of global use) of CFC's (fully halogenated chlorofluorocarbons) is to be halved by the end of 1990 as a contribution to avoid damage to the ozone layer. By the end of 1994 CFC use will be largely phased out.

The Montreal protocol has been signed.

To reduce air pollution and acidification the emissions of sulphur dioxides in Sweden will be reduced by 65 per cent in the 1980-1995 period and by 80 per cent in the 1980-2000 period. The emissions were almost halved in the period 1970-1980.

The emissions of nitrogen oxides will be reduced by 30 per cent in the 1980-1995 period.

Greater use of district heating and greater use of low sulphur fuels has gradually diminished local contamination of air.

The Protocols on reduction of sulphur dioxides and nitrogen oxides have both been signed.

Concerning marine pollutions discharges of chlorinated organic substances are to be cut by 60-70 per cent in the middle of the 1990s. The discharges at cach facility are thus expected to be reduced to max. 1.5 kg of chlorinated organic substance (TOC1) per tonne of pulp.

Discharges of nutrients are to be reduced by 50 per cent before the year 2000.

Discharges of nitrogen from municipal sewage plants are to be reduced by 50 per cent before 1995. At present the sewage water from plants covering more than two-thirds of the population is purified with chemical-biological treatment; 0.1 per cent of the population are without any treatment at all.

Agricultural leakage of nitrogen compounds is to be halved by the year 2000.

The use of fertilizers is to be reduced by 20 per cent by the year 2000.

The use of chemical pesticides by agriculture is to be halved in the period 1985-1990.

The discharges of metals, particularly mercury and cadmium, are to be reduced by 50 per cent by 1995.

Both the quantity of waste and the degree of risks involved are to be reduced as a result of measures both in the production stage and at the consumer stage. Newspapers and magazines will now be kept separated by every household and recycled to industry. Aluminium cans and batteries are also collected.

The requirements for emissions of air pollutants from waste incinerators have been made considerably more stringent. The requirements for dioxin emissions are very strict. All existing waste incinerators are to be modified by the end of 1991.

Concerning vehicle exhaust the present US Federal requirements for motor vehicles will be introduced in Sweden in 1989. The 1990 US Federal requirements for trucks and buses will be introduced in 1992 (light vehicles) and 1994 (heavy vehicles) respectively.

The lead content in gasoline is restricted to 0.15 gram per litre. Lead-force gasoline is widely available.

Guidelines concerning measures to counteract noise from road traffic have been approved.

Public Response

The public concern for the environment has risen gradually in Sweden since the 1960s.

A poll by an opinion poll institute (Sifo on behalf of the Federation of Swedish Industries, 1986) shows that the environmental issue was the most important political issue at the end of 1986, enjoying priority over issues like energy supply, income taxes, military defence, etc.

A vast bloom of poisonous algae and the deaths of seals in the North Sea in 1988 made the public's concern grow even more. Roughly two-thirds of the Swedish seal population died during a couple of weeks, primarily because of a virus infection. One result of public concern for the environment is that a new political party, "the Greens", has now entered Parliament.

Another opinion poll (Sifo on behalf of the Federation of Swedish Industries, 1987) shows that top management in Swedish industry is also concerned about the environment and is taking an active interest in environmental

protection.

However, one conclusion can be drawn from the two opinion polls concerning environment protection. Top management tend to look at the good results achieved in release reduction while the public do not have access to that information and tend to focus on the remaining problems.

Industry Activities

The Board of Directors of the Federation of Swedish Industries unanimously approved an Environmental Policy Programme for Swedish Industry in October 1987. The Board not only addresses the politicians, but also member companies. There is an appeal to top management for increased commitment to environmental matters in industry:

- to engage top management;
- to develop an environmental plan for all industries;
- to look at environmental aspects at an early stage in design of products and projects;
- to ensure that manning for environment protection is adequate;
- to review carefully licences and monitoring programmes;
- to improve the operation of environmental equipment;
- to review the potential for industrial accidents and accidental releases.

As a follow-up to this programme the Federation has prepared a special Handbook for Environment Protection in Industry.

A number of companies have their own corporate environmental policy, which entails improved environmental management. There is without doubt an increased interest within industry in including environmental management as an integral part of corporate policy.

IMPLEMENTING ENVIRONMENTAL POLICIES – THE SWEDISH APPROACH

This section will focus on one of the most important environmental acts, the Environmental Protection Act, even though for instance the Act on Chemical

Products and the Act on Transportation of Dangerous Goods are also important in this context.

Licensing

One of the most important means of control in the Swedish environmental policy is the licensing system according to the Environment Protection Act (1969). The Act applies to air pollution, water pollution, noise, solid waste and other kinds of environmental disturbances.

A licence is required for construction, expansion or alteration of certain types of stationary sources (major and medium sources), while a notification will suffice for others (minor sources).

A central governmental administrative body, the Licensing Board for Environment Protection, is responsible for licensing the potentially most polluting facilities, for example pulp and paper mills, iron and steel works, refineries etc., totalling some 500 industrial facilities (major sources). The 24 County Administrations are responsible for licensing some 4000 industrial facilities (medium sources). The counties also have notification authority.

The Environment Protection Act gives the general framework for licensing. The basic criterion is that disturbances shall be prevented as far as feasible. Environmental protection measures shall be considered on the basis of what is technically feasible using the best available control technology. The basic overall requirement is, in short, that the Authorities in their decisions shall find a proper balance between what is:

- technically feasible;
- ecologically required; and
- economically reasonable and
- ecologically required.

Thus the approach is that the authority applies the standards to polluting sources through licences which are tailored to individual facilities, taking into account different aspects as mentioned above. In each licence a number of standards for air pollution, water pollution, etc. and other

conditions are specified. The standards and conditions are generally expressed in terms of emissions per unit of time or per unit of product, concentration, etc. The decision process includes public involvement, written comments from authorities, organizations, companies and the public as well as negotiations with the company management in question.

One major advantage of individual licence conditions is that the industry is assured that it should be able to meet the requirements, even if they are tough.

During the 1970s and the beginning of the 1980s the authorities gave priority to as stingent requirements as possible before supervision and follow-up. During the last couple of years priority has been given to more effective follow-up of the licences.

Supervision and Inspection

The National Environment Protection Board is responsible for central supervision, the County Administrations for inspection of the facilities. The County Administrations may delegate this responsibility to the municipalities' Environment and Health Protection Boards. A total of some 100 inspectors are at present involved in the supervisory work. This means that on average one inspector covers five major and 40 medium industrial sources. The number of inspectors is expected to be doubled in a short time as a consequence of Parliament's environmental policy decision in June 1988. The responsibilities of the local Boards will also be expanded, a fact which will increase the number of inspectors further.

Supervision means that the Inspection Authority shall:

- verify that the licences are adequate;
- verify compliance with the regulations in the licences or with the general requirements in the legislation to protect the environment;
- check up that cases of non-compliance will be corrected, take enforcement action and, through the judicial system, prosecute violators;

– supervise the state of the environment.

The Supervisory Authorities have certain legal powers at their disposal:
– polluting activities may be prohibited;
– protective action may be prescribed, possibly under the penalty of fines;
– punishment may be by fines or up to two years' imprisonment;
– punishment may be by fines or up to six years' imprisonment in grave cases, according to the Penal Code;
– the Environment Protection Fee is to be charged for violations that have entailed economic advantage.

A fundamental principle in the supervisory system is that the licensed sources primarily shall keep track of their own compliance status based on Source Self-Monitoring. A Monitoring Programme confirmed by the Supervisory Authority specifies the required monitoring. The Monitoring Programme also often includes monitoring environmental status in the vinicity of the facility. The results of the monitoring are regularly reported to the Authority (annually, quarterly or monthly). One main reason for adopting the Source Self-Monitoring system is that management should take responsibility for all its activities, including adverse effects on the environment. Another reason is that it is probably impossible for the Authorities to get enough resources to maintain sufficient follow-up. Collecting, recording and reporting monitoring data to the Authority will also probably result in a higher level of management attention.

The approach in the legislation to individual licences presents a difficult challenge to the Inspection Authority. The Inspector needs to understand the licence for each plant before conducting an inspection. In case of non-compliance he also has to collect evidence that differs from case to case.

The Inspection Authority can also require an independent third party inspection. This is generally the case for new sources and for expanded existing sources, as well as on a routine basis for major sources. These inspections are generally conducted by a consultant and the findings are reported to the Inspection Authority.

ENVIRONMENTAL AUDITING – A NEW IDEA

A Swedish Proposal on Environmental Auditing

A Governmental Committee proposed in 1987 that about 4000 plants (strictly those plants for which an environmental protection licence is required from a governmental administrative body) in Sweden should submit an annual "Environmental Report" to the Inspection Authority. The proposal on reporting should be viewed as a confirmation of already established routines. This proposal will be implemented in 1989.

The Committee also proposed that the Environmental Report for the 300-400 biggest plants – and in the long-term perspective 4000 – should include a written review by an environmental auditor certifying that the information contained in the report is correct.

A prerequisite for the Committee's work was that the proposals should not lead to any expansion of the number of the Authorities' personnel.

In his review the environmental auditor should state his views on how far the information in the Environmental Report complies with the monitoring programme, and with the standards given in the environmental protection licence. The environmental auditor should also report on whether the facility has failed to report deviations from requirements due to accidents, operational disruptions, etc.

The information in the company's Environmental Report and the environmental Auditor's Statement should form the basis for the Inspection Authority's assessment of whether the company has fulfilled prescribed monitoring and complied with the regulations. Whenever applicable, the information should also form the basis for the Authority's decision on environmental protection requirements and on penalties.

The Committee emphasizes in its report that the introduction of environmental auditors does not mean that those who are involved in potentially environmentally hazardous operations in any way are free from their responsibility for active environmental protection. The auditors' work, however, is expected to reduce the number of violations.

The Committee also emphasizes that responsibility for inspection of environmentally hazardous operations and their impact on the environment will be retained by the Inspection Authorities. The auditors' efforts, however, are expected to facilitate the work of the Authorities.

The environmental auditor's position is expected to be independent of the companies.

The auditor is also expected to report continuously to the company management on observations and findings. These reports serve primarily to bring about corrective actions within the company.

According to the Committee the environmental auditor is qualified to advise the company concerning technical environemental protection measures. Thus the environmental auditor should play a role in the company's work to prevent disturbances and to develop environmental protection.

This part of the Committee's report was not included in the Governmental Environmental Bill to Parliament in June 1988. One reason for this was the severe criticism of the Committee's proposal from industry and the Inspection Authorities. The role of the auditor was not clearly defined. In some respects the proposed environmental auditor was both an internal industry auditor and an Authority Inspector. Thus these measures have, so far, not been implemented.

It should also be mentioned that at the time the Committee was meeting there was no information available on the environmental auditing programmes, even though such programmes had already been introduced in a number of companies in other countries.

After closer consideration the Government found that it was possible to expand the number of the Authorities' personnel.

International Environmental Auditing

During the last ten years, environmental auditing or assessment programmes have been developed in a number of US companies. These programmes have also been adopted in a number of companies in Canada and in Europe (the Netherlands, the Federal Republic of Germany). The International Chamber of Commerce has developed a position paper on this subject.

The mutual characteristics of these programmes are:
- systematic, independent evaluation of companies' environmental activities to verify at least compliance with requirements in the legislation;
- an internal management tool;
- reporting remains within the company.

In the US as well as in Canada and Sweden a discussion has taken place as to whether governments should require mandatory audit programmes.

The US and the Canadian Governments and environmental agencies have abandoned such plans because of local and industrial concern. Instead, public efforts are limited to encouraging industry to adopt such programmes by itself.

CONCLUSIONS

The most important tool the company can use to comply with the regulations is to implement a strong environment protection management programme. Good management in a company involves good environmental management. In fact many companies have adopted environmental programmes that go beyond the requirements of the legislation.

A number of people will be involved

in such a programme – from top management down to plant-floor level. The responsibility for a successful outcome of the environmental work is thus spread all over the company. It is essential for progress that the top management of the company gives a clear and strong support to the programme. It is also essential to train the staff in their tasks and supply them with adequate education and information.

It must be said, however, that it is almost an impossible task to achieve compliance for every requirement in every plant at every moment. No guidance exists indicating what the reasonable expectations should be. It is possible, however, to improve compliance and to show more clearly the intention of industry to comply with regulation.

The present rate of compliance varies. According to a systematic inspection programme in progress covering different industrial sectors, led by the National Environment Protection Board, the compliance rate among industrial plants can be estimated at roughly 90 per cent. Four industries have been investigated: iron and steel, metal finishing, fibre board and pulp. The results of the study so far show that at least one requirement had been exceeded at 10 per cent of the plants inspected during the period of the investigation. The share of conditions that were exceeded could be estimated at roughly 2 per cent. These figures roughly correspond to what has been reported from the US, for example.

In certain cases, however, better defined requirements and monitoring would have increased the number of non-compliances.

The Inspection Authorities will probably never get enough economic and personnel resources to ensure that all industrial plants are in compliance with all legislative requirements over a specific period of time. To do so, it is necessary to develop other mechanisms.

Greater emphasis on source self-monitoring and reporting, quality assurance of monitoring data, and less costly but continuous techniques for monitoring should be useful in this respect.

In addition to an effective management programme, an environmental audit or assessment programme can constitute an effective complement if used as an internal management tool.

One major advantage is that such programmes highlight industry's environmental concern.

Companies that have not shown environmental awareness and responsibility, because they have inadequate resources and know-how, cannot be forced to adopt audit programmes. Therefore, given limited resources, governmental agencies could concentrate their inspection programmes on those companies. Also the "good example" as mentioned above might make those companies change their attitudes.

The government's means of control to achieve compliance with regulations should not change as a consequence of industrial environmental auditing programming. Thus authority supervision and inspections, criminal and economic sanctions should remain as before, maybe with slightly modified priorities. To be specific, governmental supervision and sanctions are a prerequisite for effective internal industrial environmental protection programmes. However, environmental auditing can be an important tool in the efficient management of environmental work in industry and also a response to public request for greater industry involvement in environmental protection work.

CHAPTER 9 - ENVIRONMENTAL AUDITING AT CIBA-GEIGY

Dr. G. EIGENMANN
Ciba-Geigy Ltd

An overall concept of environmental protection must include proper management methods. Environmental protection must be given the same attention and care as any other field of operation. Environmental protection is an important aspect of the long range strategy of an enterprise.

Management's tasks are:
— setting goals and objectives;
— defining responsibilities;
— providing the means to achieve the goals;
— carrying out controls to check on progress.

Environmental auditing serves to check on the environmental performance at a given unit. It also serves to identify open problems and to initiate corrective actions necessary to reach the objectives set by the enterprise.

In environmental protection, goals might be limited to legal compliance (e.g. goals given by the respective national and local legislation). Such goals, however, may include more: a policy may state that besides meeting legal limits, an enterprise may also want to reach stricter objectives, if it deems such stricter standards to be necessary.

At Ciba-Geigy, the environmental objectives also include, besides legal compliance, areas which are not covered by laws, such as management and proper data bases. In certain countries they also encompass environmental areas not covered by national law.

At Ciba-Geigy an environmental audit was institutionalized in 1981, to our knowledge one of the first environmental audits within a European Company. The audit covers plants in Switzerland, as in the other group companies with the exception of the US where for legal reasons the audit is adapted to the situation of the US and is carried out by local teams.

The environmental audit at Ciba-Geigy thus covers more than legal compliance. Focal points are:
— legal compliance;
— compliance with Ciba-Geigy internal norms;
— proper environmental management;
— inclusion of environmental matters in the planning procedures;
— adequate data bases on environmental aspects;
— efficiency of pollution control measures.

The audit teams are organized independently, i.e. they do not report to the plants.

An audit is always carried out with the environmental officer of the plant being audited. Discussions are directly with staff and management of production, laboratories, shops, etc. Generally a team of two auditors works together for roughly two weeks at a site.

Generally, the audit at a plant comprises fact-finding discussions; it is closed with a meeting between the plant management and the audit team. At this meeting, the findings of the audit are discussed and counter-measures are agreed upon. The final report includes all findings, prioritized to show important deviations. The report goes to top management and forms the basis for corrective actions.

We have also installed an audit follow-up system to keep track of the status of all audits carried out so far.

CIBA-GEIGY

AUDIT PLANNING

Plan for next year

Notification of plant site
Basic information needed
Procedure and contacts
Time schedule

Fact-Finding Period
(Single time block or several short periods)
Meeting with plant management
Discussions as needed
→ Draft of provisional report
Final meeting with plant management

Reporting
Draft final report
Comments by plant:
check for factural errors
assign responsibilities
assign completion times
Issue final report
to management of Group Company and of plant site
to management at Corporate Headquarters

Audit Follow-Up

Keep track of completion of agreed-on measures

CIBA-GEIGY

AUDITING FOCAL POINTS

Management
 Organization
 Responsibilities
 Information
 Instruction

Environmental Service
 Responsibilities
 Integration in plant
 Role in new projects
 Staffing and capacity
 Documentation

Production Departments
 Responsibilities/E.P.
 Process documentation
 (air, water, wastes)
 Cost allocation
 Modern production procedures

Compliance with
 Laws
 Specific permits
 Ciba-Ceigy internal norms
 → air, water, waste, noise
 → infrastructure, production, laboratories

Pollution Control Technology and Infrastructure
 Efficiency
 State-of-the-art
 Maintenance
 Operating procedures
 → air, water, wastes

Impact on Surroundings of Plant
 Receiving rivers
 Groundwater (old landfills)
 Atmosphere (odour)

Accident Situations
 Risk of accidental emissions
 → air, water
 Preventive measures
 Information and reporting

Management of Special Wastes
 "Cradle-to-grave" controls
 Control and management systems
 Record keeping
 also:
 Auditing external disposal facilities.

INDUSTRY - ENVIRONMENT

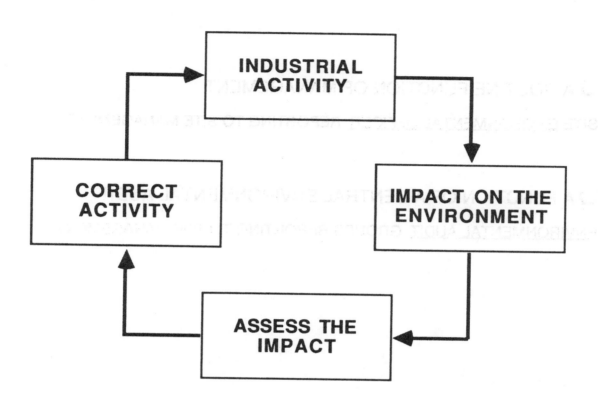

❏ ASSESS TYPE AND MAGNITUDE OF THE IMPACT ON THE ENVIRONNMENT

❏ EVALUATE ACCEPTABILITY (SHORT TERM / LONG TERM)

ASSESSMENT OF THE ENVIRONMENTAL IMPACT

❑ A ROUTINE FUNCTION OF MANAGEMENT

SITE <u>ENVIRONMENTAL OFFICER</u> REPORTING TO SITE MANAGEMENT

❑ A FUNCTION OF A CENTRAL ENVIRONMENTAL AUDIT

<u>ENVIRONMENTAL AUDIT</u> GROUPS REPORTING TO TOP MANAGEMENT

OPPORTUNITIES THROUGH ENVIRONMENTAL AUDITS

❏ ASSURE COMPLIANCE WITH STANDARDS

❏ ASSURE EFFICIENT TECHNOLOGY

❏ AVOID PROBLEMS
   SHORT TERM / LONG TERM

❏ PROBLEM SOLVING THROUGH INFORMATION TRANSFER

> AN ENVIRONMENTALLY ACCEPTABLE
> OPERATION IS AN IMPORTANT LONG
> TERM MANAGEMENT OBJECTIVE

ENVIRONMENTAL AUDITING

COMPARING:

❑ OBJECTIVES/STANDARDS
        WITH
❑ ACTUAL PERFORMANCE

AUDIT AREAS MAY BE:

❑ LEGAL STANDARDS

❑ INTERNAL NORMS
    - TECHNICAL
    - MANAGEMENT

❑ LONG TERM PLANNING ASPECTS

STEPS OF AN ENVIRONMENTAL AUDIT

❏ PREPARATION AT THE SITE
❏ FACT FINDING, DISCUSSIONS ON SITE
❏ EVALUATION OF FINDINGS
❏ ASSIGNMENT OF RESPONSIBILITIES & DEADLINES
❏ REPORTING
❏ FOLLOW–UP

AUDIT REPORTING

A REPORT MUST BE :

❑ BASED ON FACTS

❑ UNPERSONAL

❑ THE REPORT SHOULD BE ACCEPTED BY THE SITE MANAGEMENT

❑ IT MUST INCLUDE AGREED-ON CORRECTIVE ACTIONS
- RESPONSIBILITIES
- DEADLINES

AREAS OF CONCERN

❑ <u>STANDARDS NOT CLEARLY DEFINED</u>
   INTERNAL / LEGAL

❑ <u>LEGAL ASPECTS</u>:
   WRITTEN EVIDENCE OF NON - COMPLIANCE

❑ TECHNICAL COMPETENCE
   OF AUDITORS

THE AUDITORS ROLE IS NOT THAT OF AN ENVIRONMENTAL POLICEMAN

CHAPTER 10 - NORANDA INC.'S ENVIRONMENTAL AUDITING PROGRAM

Hennie VELDHUIZEN
Noranda Inc.

1. INTRODUCTION

Noranda Inc. is an international, diversified natural resource company with more than 100 operating facilities in four main business units — minerals, forest products, energy and manufacturing.

In 1985, the Board of Directors approved an Environmental Policy which requires that "Noranda operations will strive to be exemplary leaders in environmental management by minimizing the environmental impact on the public, employees, customers and property". In addition, it dictates that "Noranda Group operations will be subject to periodic environmental, health, hygiene, safety and emergency preparedness audits".

The auditing program is an extremely useful management tool which enables Noranda to be an exemplary leader in environmental management.

This paper provides an overview of what Noranda Inc. is, the changing environment in which we operate, the Environmental policy statement and the environmental auditing program being implemented within this large diversified company.

2. NORANDA INC. – A VERY RESOURCEFUL COMPANY

Noranda Inc. started as a producer of copper and gold in 1928, but through diversification now operates in the entire spectrum of natural resources. It is an international company with combined annual sales of over $ 8 billion.

Noranda Minerals Inc. is a major producer of: copper, leadf, zinc and related by-products gold and silver; molybdenum; sulphuric acid, potash and phosphate fertilizer. Products are marketed in 65 countries.

Noranda Forest Inc. was formed in 1987 as Canada's largest forest products group. It has interest and markets around the world. The Company owns or controls large timberland areas, and has timber-cutting rights across Canada and the United States which provide many of the raw materials for its three principal businesses — pulp and paper, building materials and paperboard packaging. It is currently developing a $ 1 billion pulp mill in Tasmania with an Australian partner.

Noranda Energy is comprised of three innovative companies which rank it among the top ten in Canada in terms of oil and gas reserves. Its holdings include interest in Canada's Beaufort Sea, the United States Gulf Coast, California, Australia and New Zealand.

Noranda Manufacturing encompasses 28 manufacturing plants and 230 sales and distribution centres throughout Canada and the United States. This group's facilities in North America and overseas produce a diversified range of materials including aluminum metal, fabricated aluminum building, sheet and foil products and steel wire rope and cables.

Accountability and responsibility for

management is very much decentralized to the units within each business element of the corporation. Individual plant managers are fully accountable for: maintaining and improving productivity, efficiency and competitiveness; reducing operating costs; protecting the environment, worker health and safety, and the public with respect to any aspect of plant operations; and maintaining credibility and good relations with governmental agencies, the public, employees and customers.

The Company employs 44.000 men and women world-wide in plants, sales offices and distribution centres ranging from less than 50 to more than 1000 personnel. Consequently, the number of staff members supporting the environmental program at each operating facility varies from less than one to perhaps a half dozen.

3. THE ESCALATING ENVIRONMENTAL REVOLUTION

Public support for environmental protection has never been at a higher level. Opinion polls, from the early 1980's to the present, have clearly demonstrated that 70-85 per cent of the Canadian population consider the environment and the protection thereof to be an extremely important issue. Global and local environmental events, featured almost daily on television and radio and in the print media, ensure a well informed and an aware public — a public which is now demanding of its legislators and its industry a higher degree of environmental protection for this and future generations.

The message has not been lost on federal, provincial and local politicians. All parties have adopted responsible environmental management as a key campaign issue and one emerging party — the Greens — champions environmental protection as its modus operandi. Politicians recognize that a strong environmental commitment is essential for re-election.

A series of relatively recent environmental incidents and longer term potential impacts associated with industrialization — transportation accidents, leaking landfill sites, improper hazardous waste disposal, contaminated surface and sub-surface drinking water supplies, health concerns allegedly related to pollutant exposure and acid rain, global warming and ozone depletion — have served to focus public attention on the inadequacy of current environmental management systems.

In response, an ever increasing volume of governmental regulations, acts, guidelines and codes of practice have been formulated over the past ten years and will continue to be developed in the next decade. Not only are these regulatory instruments becoming more comprehensive and more complex but they are also incorporating innovative enforcement techniques and more substantive penalties. Incarceration of chief executive officers and managers for up to several years and fines extending into the million dollar range are now common.

Liabilities associated with past hazardous waste management and disposal practices, product and industrial exposure health claims and off-site impacts due to industrial accidents can threaten the financial security of a corporation. Noranda has concluded that it must advance its environmental management systems for the protection of the environment, its employees, the public and its financial assets.

How does Noranda Inc. propose to accomplish this? How does it assist more than 100 plant managers from small to large facilities in the advancement of the environmental program?

4. THE NORANDA INC. ENVIRONMENTAL POLICY

A clear corporate position is essential for a consistent environmental management system throughout such a broadly based and diverse company. In 1986, the Board of Directors approved the following Environmental Policy statement.

"Noranda operations will strive to be exemplary leaders in environmental management by minimizing the environmental impact on the public, employees, customers and property, limited only by the technological and economic viability. The following principles are basic to achieving this environmental objective.

1. The potential risks of new projects to employees, the public and the environment must be assessed so that effective control measures can be foreseen and taken and all parties made aware of these facts.

2. Noranda Group operations will implement site specific environmental, health, hygiene, safety and emergency response policies in the spirit of guidelines issued by Noranda Inc. as well as in conformation with applicable laws and regulations.

3. Noranda Group operations will constantly evaluate and manage risks to human health, the environment and physical property.

4. Noranda Group operations will be subject to periodic environmental, health, hygiene, safety and emergency preparedness audits.

5. A report on environment, health and hygiene, safety and emergency preparedness will be presented annually to the Board".

The policy, signed by Alfred Powis, Chairman and Chief Executive Officer, is not just a statement intended for senior management within the corporation but rather a clear message for all 44.000 employees who individually are essential for an effective environmental program. It enunciates the Board's commitment to and support for proactive environmental management. It expects each employee to make the environment part of his or her business.

It should be noted at this point that, by Noranda Inc.'s definition, environment encompasses the external environment, occupational health, industrial hygiene, emergency preparedness and safety.

Principle four requires periodic environmental, health, hygiene, safety and emergency preparedness audits. Principle five, which is an extremely important one, closes the accountability loop by specifying the submission of annual "state-of-the-environment" reports to the Boards; there is a reporting back to the Directors on how the corporation is doing in environmental compliance.

5. THE ENVIRONMENTAL AUDITING PROGRAM

5.1 Definitions, Objectives and Goals

Environmental auditing is a systematic, objective method of verifying that environmental, health, industrial hygiene, safety and emergency preparedness standards, regulations, procedures and corporate guidelines are being followed. The examination involves analysis, testing and confirmation of procedures and practices at an operation. In addition, the process also evaluates the adequacy of the environmental management system — communications, delineation of clear employee responsibilities, training, quality control — the risk assessment process and the application of best management practices at facilities.

Although the auditing program developed and introduced at Noranda in 1985 is still an evolving and dynamic system, some general objectives have been developed including compliance with regulatory requirements and Noranda guidelines, application of best management practices, minimization of potential risks and liabilities, and application of good management systems to the environmental program.

In support of these objectives, Noranda Inc. has established specific goals which are to:

— audit all operations at least once every four years;

— correct all deficiencies and findings in a timely and cost–effective manner;

— reduce liabilities and risks to a minimum by correcting previous practices, by improving engineering designs, by process modifications and by chemical substitution;

- improve awareness and understanding of environmental regulations, standards, guidelines and codes of practice among Noranda operational staff at all facilities;
- transfer technology and an improved awareness of good environmental management systems among Noranda plants; and
- improve the efficiency and the cost-effectiveness of the environmental program.

5.2 The Auditing Process

The Noranda program is structured on well-developed and accepted audit procedures currently applied in industry. Basic steps of a typical audit outlined in Figure 1 include:

- developing an understanding of the plant's internal management systems and controls;
- assessing the strengths and weaknesses;
- gathering audit evidence through assessment and verification;
- evaluating audit findings;
- discussing findings with facility management;
- preparing audit findings for the close out meeting;
- preparing the draft audit report followed by the final version;
- completing the action plan (by the audited facility); and
- ensuring that the action plan has been implemented. This is accomplished by the corporate Environmental Services staff including the Manager of Environmental Audits during regular and special visits to plants.

Critical to the process are the preparation of an audit report, response plans and the follow-up activities described in Figure 2. The preparation and implementation of the action plan are essential for an effective audit program. This plan closes the loop and ensures that all deficiencies and findings are corrected in a timely and cost-effective manner. It represents the due diligence element of the program.

5.3 Audit Teams, Training and Tools

In developing its program, Noranda determined that the use of internal auditors, primarily trained specialists selected from operating facilities in each business unit, would best serve its objectives and goals. The primary reason is that the extensive knowledge, experience and environmental awareness developed during the auditing exercise remains within the corporation. In fact, the in-depth understanding of regulations and the exposure to excellent management systems and environmental controls at other plants are valuable assets which the auditors transfer to their own and other facilities.

Auditor candidates have been selected from plants within each business unit supplemented by staff from the corporate Environmental Services. The professions include chemist engineers, environmental management scientists, industrial hygienists, medical doctors and nurses, and process, transportation and emergency response experts. Extensive training for auditors has been provided by a specialized consulting firm, which in addition to teaching the techniques and process, lead the participants through a trial review. Yearly workshops are held thereafter for updating skills.

5.4 Audit Types

Seven distinct types of environmental audits have been conducted since 1986. These include the external environment, occupational health, industrial hygiene, emergency response, acquisition, divestiture and closure. Detailed and separate protocols have been developed for the first four. For acquisitions, divestitures and closures there is a greater emphasis on the identification of liabilities, the costs associated with upgrading the facility to today's standards and the issues which must be addressed in a closure or reclamation plan.

The review elements for environment, industrial hygiene, occupational health and emergency preparedness against which the audit is conducted are summarized in Tables 1 to 4. Protocols are divided into two distinct areas – compliance with federal, provincial, state and municipal regulations, standards and by-laws as well as corporate standards, policies and guidelines and best management practices. Prevention of potential environmental impacts through the application of precautionary procedures is an intentional, more cost-effective strategy than reaction.

All environment, occupational health, industrial hygiene, emergency response, acquisition and cloture audits are conducted with internal personnel. Teams of three to four auditors are assigned to a specific task by the Manager, Environmental Audits based on the scope of the review and the expertise required. For small facilities, environment, occupational health and industrial hygiene audits may be conducted at one time in which case the team will have expertise from each area. However, in the case of large plants, a review will generally be limited to one protocol requiring three or four experts in that particular field. Each auditor may be assigned to two or three audits per year. Currently, Noranda has 55 trained auditors who have completed 56 audits to the end of 1988.

Third party external auditors are used for divestiture reviews where an unbiased assessment of potential liabilities and the costs for upgrading environmental systems is required.

5.5 Audit Report Security

Final audit reports and action plans are distributed to the specific plant manager, his superior in that particular business unit, the Vice-President of Environmental Services and the Manager of Environmental Audits.

The question of report accessibility by governmental agencies and by court directed disclosure is one that industry and legislators struggle with. Noranda has concluded that the benefits flowing from the auditing program far outweigh any negative concerns associated with disclosure. A due diligence court

defence, consisting of an active audit program which critically and independently (independent to the particular plant in question), identifies and corrects environmental deficiencies and findings represents the best defence.

5.6 Priority Setting and Audit Scheduling

Acquisition, divestiture and closure audits are obviously scheduled on an as needed basis. The Manager, Environmental Audits plans a four year schedule for the environment, occupational health, industrial hygiene, product safety and emergency preparedness audits based on the input of corporate Environmental Services and the following factors:

- new or modified legislation;
- the size of the facility;
- the processes carried out and the characteristics of the chemicals and raw materials used and the volumes stored;
- the employee exposure to in-plant chemicals and process by-products;
- the emission, effluent and waste volumes and characteristics;
- the sensitivy of the environment surrounding the facility and the nature of the receiving environment; and
- the proximity of public residences to the plant.

Although the intent is to audit each facility once every four years, operations which have a higher rating in terms of the above noted criteria are scheduled more frequently.

6. CONCLUSIONS

Noranda's objective is to be the premier diversified natural resource company in the world. Some resources are dependent on the quality of the environment (forestry) and others through their development have the potential to impact on the environment. The Environmental Policy dictates that "Noranda operations will strive to be exemplary leaders in environmental management by minimizing the environmental impact on the public, employees, customers and property".

Environmental auditing is an extremely important management tool for Noranda — one which supplements the diligent activities of plant personnel and plant managers and the regular compliance inspections and evaluations carried out by the corporate staff of Environmental Services. It enables the corporation to fulfil its mandate to demonstrate exemplary leadership. It moves the company forward beyond compliance requirements to an upgrading of its facilities in terms of best management practices. It is the best

due diligence defence Noranda can offer.

Good environmental management has the full commitment and support of the Noranda Board of Directors, its management and employees. Environmental auditing and the Policy statement are a clear demonstration of the corporation's intention and commitment to exemplary leadership in environmental management. They lend encouragement to each employee to make the environment an integral part of his/her business.

Figure 1

BASIC STEPS OF AN ENVIRONMENTAL AUDIT

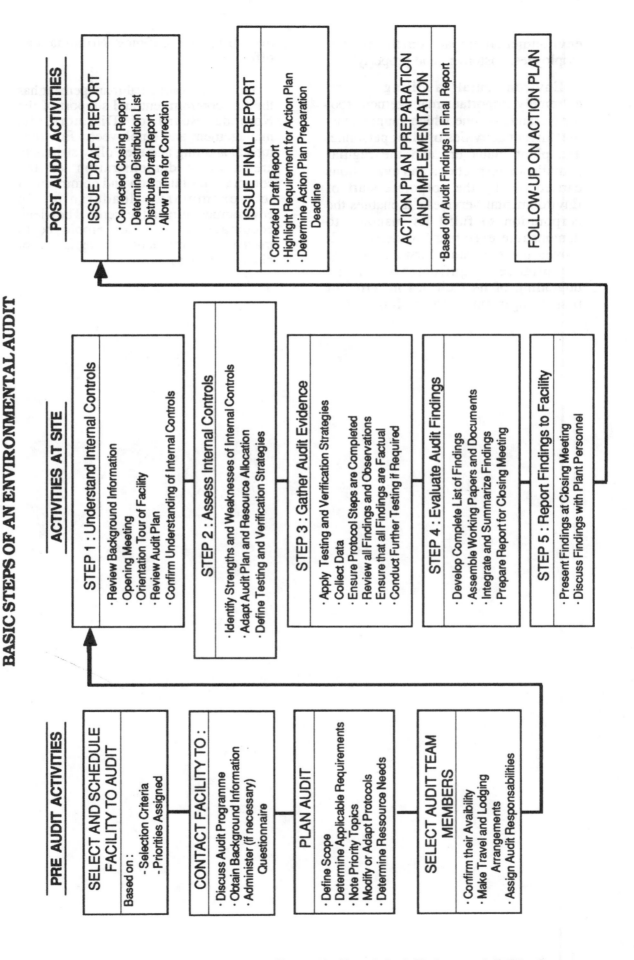

PRE AUDIT ACTIVITIES

SELECT AND SCHEDULE FACILITY TO AUDIT

Based on :
- Selection Criteria
- Priorities Assigned

CONTACT FACILITY TO :
· Discuss Audit Programme
· Obtain Background Information
· Administer (if necessary) Questionnaire

PLAN AUDIT
· Define Scope
· Determine Applicable Requirements
· Note Priority Topics
· Modify or Adapt Protocols
· Determine Ressource Needs

SELECT AUDIT TEAM MEMBERS
· Confirm their Avaibility
· Make Travel and Lodging Arrangements
· Assign Audit Respsonsabilities

ACTIVITIES AT SITE

STEP 1 : Understand Internal Controls
· Review Background Information
· Opening Meeting
· Orientation Tour of Facility
· Review Audit Plan
· Confirm Understanding of Internal Controls

STEP 2 : Assess Internal Controls
· Identify Strengths and Weaknesses of Internal Controls
· Adapt Audit Plan and Resource Allocation
· Define Testing and Verification Strategies

STEP 3 : Gather Audit Evidence
· Apply Testing and Verification Strategies
· Collect Data
· Ensure Protocol Steps are Completed
· Review all Findings and Observations
· Ensure that all Findings are Factual
· Conduct Further Testing if Required

STEP 4 : Evaluate Audit Findings
· Develop Complete List of Findings
· Assemble Working Papers and Documents
· Integrate and Summarize Findings
· Prepare Report for Closing Meeting

STEP 5 : Report Findings to Facility
· Present Findings at Closing Meeting
· Discuss Findings with Plant Personnel

POST AUDIT ACTIVITIES

ISSUE DRAFT REPORT
· Corrected Closing Report
· Determine Distribution List
· Distribute Draft Report
· Allow Time for Correction

ISSUE FINAL REPORT
· Corrected Draft Report
· Highlight Requirement for Action Plan
· Determine Action Plan Preparation Deadline

ACTION PLAN PREPARATION AND IMPLEMENTATION
· Based on Audit Findings in Final Report

FOLLOW-UP ON ACTION PLAN

Figure 2

POST REVIEW ACTIVITIES

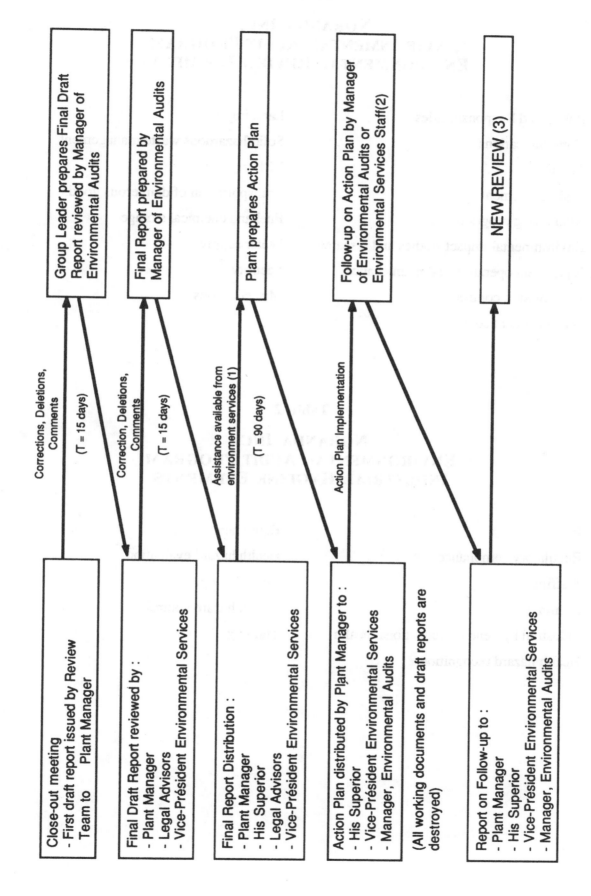

Table 1

NORANDA INC.
ENVIRONMENTAL AUDIT PROGRAM
ENVIRONMENTAL REVIEW ELEMENTS

Policy and Responsibilities

Communications

Training

Risk assessment

Monitoring program

Environmental impact studies and research

Equipment operation and maintenance

Air emission control

Water effluent control

Land impact

Solid/hazardous waste management

PCB program

Transportation of dangerous goods

Fuel and chemical storage

Water supply

Security

Miscellaneous

Table 2

NORANDA INC.
ENVIRONMENTAL AUDIT PROGRAM
INDUSTRIAL HYGIENE ELEMENTS

Policy

Regulatory compliance

Staffing

Facilities

Industrial hygiene / Occupational health

Health hazard recognition

Equipment

Health hazard evaluation

Record-keeping

Health hazard control

Training

Table 3

NORANDA INC.
ENVIRONMENTAL AUDIT PROGRAM
OCCUPATIONAL HEALTH ELEMENTS

General

Medical surveillance

Workplace contamination

Medical surveillance - Jobs

Programs

Company health department

Company health relation with other occupational health specialists

Meetings

Equipment

Table 4

NORANDA INC.
ENVIRONMENTAL AUDIT PROGRAM
EMERGENCY PREPAREDNESS

Policy and responsibilities

Risk evaluation

Emergency plan framework

Emergency equipment and physical facilities

Internal communications

Training

Community response and public relations

Legal

Security

Plant maintenance

Miscellaneous

CHAPTER 11 - THE TASKS OF THE CORPORATE ENVIRONMENT OFFICE IN GERMANY

Dr Egon KELLER
Ecosystem

Tasks of the Corporate Environmental Officer

Annual Report to Top Management

Solid Waste

Survey by State
Authority (biannual)

Hazardous Waste
Documents

Wastewater and Sewage

Survey by State
Authority (biannual)

Waste Oil Documents

Emissions

Emissions Report
annual 31.05 to Local/Regional
Industrial Regulatory Agency

Report on Installation Changes
biannual to every Local/Regional
Regulatory Agency

Legal Requirements for Production Plants

Environmental Protection

Solid Waste

Federal Waste Disposal Act

Wastewater and Sewage

Federal Water Management Act

Groundwater Ordinance

Federal Wastewater Act

Act on Hazardous liquids

Federal Used Oil Act

Emissions

Federal Environmental Quality Act

Technical Guideline on Noise Abatement
Technical Guideline on Clean Air

Authority in Charge

Solid Waste

Municipal Authority

City Office
County Office
Municipal CleaningServices

**Industrial Waste =
Domestic Waste**

Waste Disposal Sites

Incineration Plants

Hazardous Solid Waste

Recycling
Authorized waste disposal land sites

Incineration (corporate or public)
Incineration at sea
Open sea dumping

Underground disposal

Wastewater and Sewage

Lower Water Authority

Engineer's Office
Waste Water Cooperative (responsible)
Lower Water Authority

**Industrial Waste Water =
Domestic Waste Water**

Sewage system

Disposal in rivers

Hazardous Liquids

Recycling
Sludge, Filtration
Treatment
Emulsion Cracking

Evaporation
Acid Base Neutralisation
Galvanic Liquid Detoxification

After treatment : liquid and sludge

Emissions

Industrial Regulatory Agency

Dust, Gases, Noise

Release into Atmosphere

Emissions Standards

Dust : Dry Precipitator
        Wet Precipitator
        Electric Precipitator

Gases : Scrubber
        Electric Precipitator with Product
        Liquid and Sludge

Noise : Noise Protection Walls
        Insulated Windows
        Abatement at Source

CHAPTER 12 - QUALITY CONTROL IN THE ENVIRONMENTAL PROTECTION POLICY OF THE RHONE-POULENC GROUP

Jacques SALAMITOU
Rhône-Poulenc

A lot of attention has been focussed on the "environmental audit" as the method of assessment of environmental protection in industrial facilities.

In its now internationally accepted meaning, mostly through the influence of the North American Companies who were most advanced in those matters, the environmental audit has been defined as a method of assessing environmental performance in a way similar to the one, which *mutatis mutandis* has been extensively used to check up on the quality and correctness of the accounting activities of a firm. Since the accounting audit is justified by the possibility that some unscrupulous accountant might cheat either the company, its shareholders or the tax agency, an environmental audit may also be perceived purely as a way of discovering "cheaters" in the area of environmental protection. Since environmental auditing is and should be performed on a voluntary basis, too great a focus on the word audit may have a very negative drawback in terms of its efficient introduction in companies operating in countries whose psychology is different from the North American one.

Environmental auditing is certainly a great tool for the planning and direction of environmental protection, but it is necessary to consider it only as a part, if no doubt a major one, of management policy. Besides, sometimes it may be have different meanings; that is why this presentation will consider the overall aspect of quality control in environmental protection.

At Rhône-Poulenc, environmental protection relies on three major points:

1. Waste minimization through development of clean technologies and recylcing of by-products;
2. Responsible management of waste elimination and control of accidental effects;
3. Development of communication both internally with the company's employees and externally with the general public.

Good quality control should consider proper implementation of these three points in all operations performed in the name of R-P.

This environmental protection policy depends heavily upon an efficient prevention system.

NEW INVESTMENTS

We consider that it is much easier to improve the quality of the environment of facilities by making sure that all new investments include all the adequate devices to reduce pollution, in agreement with general policy.

This is why R-P'S President has made it mandatory that all new investments must be reviewed by a special team belonging to the "Délégation à l'Environnement". The "Délégation à l'Environnement", part of the General Directorate for Quality, Safety and Environment, is responsible for all environmental matters in all the R-P operations world-wide and reports to the Executive Vice-President.

In fact, no investment can be approved either by the General Manager of the Division (there are seven Divisions in R-P) or the Executive Vice-President of the Group, depending on the amount, if an environmental audit has not been performed by the Délégation à l'Environnement.

So in actual practice, at Rhône-Poulenc, "environment audit" commonly refers to the review conducted before the final approval of any investment project.

It should be noted that all projects are submitted to that review, whatever their capital amount or their apparent potential impact on the environment. It is up to the Délégation à l'Environnement to decide upon the actual impact, not the operational unit.

This practice has been in operation for several years, and the initial reluctance shown by some divisions on the argument that it would delay final approval of major investments has disappeared. First of all, the Délégation à l'Environnement has always performed in such a way that no real approval delay could be attributed to the environmental audit process. And operating divisions have discovered the advantages of preventing over curing, even in the case of environmental protection.

This review is performed once the investment is finalized; but more and more frequently the operating units consult the Délégation à l'Environnement at an early stage of the design process because they do not want their project to be turned down at the end or to be asked for too many modifications. This shows the acceptance of the environmental audit but it also increases the work of the Délégation.

In fact, very few investment projects have been completely turned down. The simple fact that the designers know that their project will finally be reviewed from the environmental point of view makes them more sensitive from the beginning. But most final reviews do contain remarks on some points, and final approval is only given provided that these remarks are taken into consideration in the ultimate design of the project.

In 1988, 160 such environmental audits were conducted by the Délégation à l'Environnement. They were performed with reference to the existing regulations, but also to internal company standards, the personnal judgment of the person who conducted the audit, and discussions with the designers and future plant operators after the visit.

The environmental audit for new investments includes both chronic and accidental aspects; it is conducted together with a "safety audit" concerned with protection of personnel.

EXISTING OPERATIONS

As a preliminary, it is important to realize that the environmental audit for new projects is very often the perfect occasion to review the environmental compliance of the existing surrounding equipment with which the new investment will be incorporated. Sometimes the report for the new investment contains observations on the surrounding facilities upon which final approval depend. In addition, three kinds of assessment reviews are conducted for existing operations:

- **Audits for specific facilities**: these are conducted with reference to internal norms for the facilities: i.e. warehouses, chlorine tanks.
  A team of two is sent to the premises for that purpose. At the end of their visit, there is a meeting with the site manager in order to discuss immediate implementation of measures if some situations are judged undesirable.
  A complete report is written in the days which follows containing a complete list of requests and a calendar for implementation of the measures.
- **Safety review of productions plants**:
  This kind of review is mandatory for installations listed in the Directive but they are also conducted on similar installations in countries outside and on installations not listed in the Directive world-wide.

The review is conducted in accordance with an original R-P method by a multi-disciplinary team (engineering, environmental protection, safety engineers responsible for operations at other sites). Sheets are produced for all observed situations. These sheets are rated 1 to 3 according to the overall judgment of the team, 1 being given to undesirable situations requiring immediate correction; 2 to situations to be changed in the near future; and 3 to acceptable situations. The corrections may include new procedures or new investments.

The length of the review is from two to ten days, depending upon the installation. Preliminary conclusions are also given to the site manager at the end of the visit.

- **Visits to operating sites:** Periodically, R-P operating sites are visited by teams from the Délégation à l'Environnement to assess the actual status and practice of environmental protection. All aspects of environmental protection both chronic and accidental, are reviewed according to a check-list. The environmental plan of the plant is reviewed on this occasion and newly installed equipment in particular is visited and controlled.

Items discussed include: aqueous and gaseous emissions; compliance with permits; analysis of possible releases; underground waters; waste treatment; outside contractors; warehouses, emergency plan; training programs; expenses and investments.

This visit could be considered as close to the "environmental audit" described in several documents, but it is not intended to have usual characteristics of an audit.

In conclusion, quality control in environmental protection policy must be very efficient and not only because it involves annual costs of over 1.5 billion francs at R-P. It includes several tools. The efficiency of these tools depends very much upon the quality of the relationships which are established between the Délégation à l'Environnement and the managers responsible for operating plants. More and more, we are happy to say the latter require from us the use of the these tools.

CHAPTER 13 - VIEWS ON ENVIRONMENTAL AUDITING : A WRITTEN CONTRIBUTION FROM THE CONFEDERATION OF FINNISH INDUSTRIES

Dr. E. TOMMILA
Confederation of Finnish Industries

The large multinational corporations, with facilities throughout the world, were the first to implement environmental auditing when they wanted to check the performance of environmental management in different facilities. The aim has been to assure compliance with regulations and corporations' own policies and standards based on good industrial practices.

Environmental auditing, as it is widely understood, has not yet been widely applied by Finnish industry. Some companies with facilities in different countries are going to implement the procedure as a part of their internal supervision, but much practical experience is not yet available. The audit procedures have, however, been developed. They reflect, for most important topics, the features mentioned in the ICC draft position paper, which is also strongly supported by some organizations of the chemical industries like CEFIC and CMA.

The Confederation of Finnish Industries views on the following points are:

What concept(s) does the words "environmental auditing" cover? Are there different types of "environmental audits"?

What are the goals and objectives of environmental audits?

How do environmental audits fit in the overall policy of a firm?

Environmental auditing can be seen as a part of corporate environmental management which also includes policies. The policies set up the goals to achieve good environmental performance. Also programmes, procedures, guidelines and action plans inside individual operating facilities will be needed for the implementation of these goals in practice.

Environmental auditing provides for a systematic professional approach to assess how facilities are complying with the corporate goals. The main purpose is to ensure that proper procedures and systems are used to safeguard the environment and to get compliance with regulations. The performance is always assessed against requirements. Because senior persons with experience in good industrial practice are always used as auditors, this experience will be another yardstick.

The environmental auditing is based on existing data given by plant personnel and means a gathering of the representative data and evidence rather than a detailed review of all available data. A typical audit does not include investigations creating new data.

The audits are beneficial to the audited units, giving tools for self-regulation and advice for line management. Most of the benefits are coming out of the fact that because of the international character of the audit, the discussions and corrective measures can be quick and straightforward.

The objectives and the main priorities of audit programmes can vary a lot between different facilities and companies, and they always reflect the plant to be audited. The environmental auditing programmes and objectives should always be individually designed and implemented to meet the needs, policies and culture of the corporation.

Auditing is often concentrated on regulatory compliance and the management system (called "general compliance management audits"), but sometimes they can also contain some more specific areas, depending on the situation of the plant. Besides pure environmental issues the environmental audits sometimes also include health and safety aspects. In that case they are usually called "environmental, safety and health audits".

What are the elements of an environmental audit?

The essential elements in an auditing process are:

- collection of the existing relevant data and information of the unit;
- critical professional evaluation and assessment of the data using both experience and properly set requirements as yardsticks;
- reporting of the relevant findings, conclusions and recommendations of aspects needing improvement.

What are the procedures and the methodology to perform an environmental audit?

The auditing activities need some specific procedures and common methodology in order to guarantee the best productivity and uniformity from audit to audit. The audit has three distinct steps:

- pre-audit preparation;
- the on-site audit;
- reporting.

Pre-audit activities include the definition of key objectives and scope, the selection of the audit team, obtaining the background information needed, and the preparation of specific check-lists and audit plans.

Auditing at the site basically includes the relevant data gathering and assessment of the data. The data is gained by using check-lists, interviewing the key plant personnel, visiting the facilities and reviewing and checking the documentation. At the end of the on-site audit the wrap-up meeting is held, where the brief summary of main findings and conclusions is presented and discussed with plant management.

Every audit is followed by a written report, which draws attention to matters needing improvement and gives recommendations for line management compliance. It is an internal report with very limited distribution to the appropriate management of the company.

The development of corrective action plans is needed for follow-up of implementation of recommendations made in the audit report. A re-auditing after three to four years can also be needed, and that of some selected criticial areas even sooner.

Environmental audits in small firms

The successful implementation of the auditing programme within a company always requires full management support and commitment from both top and line management. First of all it means an understanding of the benefits of auditing, the allocation of competent senior manpower resources for the auditor groups, and active follow-up of the auditing and the action plans for recommended improvements.

CHAPTER 14 - INTERNATIONAL CHAMBER OF COMMERCE POSITION PAPER ON ENVIRONMENTAL AUDITING

PREAMBLE

In recent years public concern about the environment has led to much new legislation in many countries, and to numerous international Conventions and other accords. The business community recognizes the need for a regulatory framework in the environmental area and emphasizes the necessity of proper consultations at the preparatory stage.

Inherent in all International Chamber of Commerce (ICC) activities is the strong belief in maximizing the use of self-regulation by the business community in the spirit of responsible care. This belief is based on two fundamental considerations. First, if properly applied, self-regulation is frequently more effective than reliance on legislation and official regulations. Secondly, excessive proliferation of regulations is counter-productive. Legislation/regulations may date quickly and cannot cover every contingency. Attempts to enforce them systematically in every enterprise would mean devoting major resources to a large bureaucracy, a substantial cost to the tax-payer and a brake on a dynamic economy.

The ICC believes therefore that effective protection of the environment is best achieved by an appropriate combination of legislation/regulation and of policies and programmes established voluntarily by industry, as reflected in the ICC's environmental philosophy given in its Environmental Guidelines for World Industry (*). Environmental auditing is an important component of such voluntary policies. Industry is particularly well placed to develop environmental auditing within the concept of self-regulation.

This position paper's principal objectives are:

1. To establish the meaning of the term "environmental auditing", and to emphasize that auditing should be the responsibility of companies themselves;

2. To stress to management the benefits of environmental auditing as a highly desirable and cost-effective means of assessing the functioning of any enterprise from an environmental viewpoint;

3. To emphasize to relevant authorities that such audits can, inter alia, be a reliable, flexible and efficient means of assisting compliance with regulations;

4. To help to establish audits as a credible and trustworthy instrument

---

* 1986 revision available as Publication 435, issued in Dutch, English, French, German, Italian, Portuguese, Spanish and Turkish editions (others are in preparation).

in the minds of the workforce, local community, environmentalist associations and the public at large;

5. To set an agreed basis for discussion and action world-wide, for use in the numerous international organizations directly or indirectly concerned with environmental questions, and in national business organizations which look to the ICC to provide advice on policy in such areas;

6. To suggest a standard practical methodology for personnel specifically charged with undertaking environmental audits.

The paper represents input from a number of expert sources, and reflects experience in countries and companies where the practice of environmental auditing is well established.

A relatively short paper evidently cannot cover all aspects of the subject. For example, in many cases environmental audits will in practice form part of inspections which are also designed to control health and safety factors within the enterprise. Moreover, the ICC fully recognizes the need for a flexible approach. The ICC itself stands ready to prepare complementary position papers as and when necessary.

DEFINITION AND PURPOSE OF AUDITS

During recent years the concept and practice of environmental management have developed rapidly within industrial organizations. The ICC Guidelines mentioned above represent one illustration of this development. The underlying objective of environmental management is to provide a structured and comprehensive mechanism for ensuring that the activities and products of an enterprise do not cause unacceptable effects in the environment. All stages are considered from initial planning and conception to final termination.

Most approaches to environmental management include provision for systematic examination of performance to ensure compliance with requirements. This paper is concerned with such systematic examination of performance during the operational phase of the

industrial activity. The distinction is made between this process and, for example, Environmental Impact Assessment (EIA) which considers potential environmental effects during the planning phase before an operation starts. Various terms have been used for such examination (audit, appraisal, survey, surveillance, review) leading to possible confusion. Here the term "Environmental Audit" is adopted.

The advantages and nature of environmental audits are considered below. However, the broad purpose is to provide an indication to company management of how well environmental organization, systems and equipment are performing. If this purpose is to be fulfilled with full cooperation and commitment of those involved, it is essential that the procedure should be seen as the responsibility of the company itself, should be voluntary and for company use only. Thus audits would not normally be used to instigate prosecutions or litigation. Accordingly, the definition of environmental auditing adopted here is as follows:

"A management tool comprising a systematic, documented, periodic and objective evaluation of how well environmental organization, management and equipment are performing with the aim of helping to safeguard the environment by:

(i) Facilitating management control of environmental practices;

(ii) Assessing compliance with company policies, which would include meeting regulatory requirements".

ADVANTAGES OF AUDITS

The primary and obvious advantage of environmental auditing is to help safeguard the environment and to assist with and substantiate compliance with local, regional and national laws and regulations, and with company policy and standards. A related advantage is reduced exposure to litigation and regulatory risk (e.g. penalties, additional regulations). The process ensures an independent verification, identifies matters needing attention and provides timely warning to management of potential future problems.

Experience demonstrates that environmental audits can have other benefits, the importance of which may vary from situation to situation, as follows:

1. Facilitating comparison and interchange of information between operation or plants;
2. Increasing employee awareness of environmental policies and responsibilities;
3. Identifying potential cost-savings, including those resulting from waste minimization;
4. Evaluating training programmes and providing data to assist in training personnel;
5. Providing an information base for use is emergencies and evaluating the effectiveness of emergency response arrangements;
6. Assuring an adequate, up-to-date environmental data base for internal management awareness and decision-making in relation to plant modifications, new plans, etc.;
7. Enabling management to give credit for good environmental performance;
8. Helping to assist relations with authorities by convincing them that complete and effective audits are being undertaken, by informing them of the type of procedure adopted;
9. Facilitating the obtaining of insurance coverage for environmental impairment liability.

It should be emphasized again that the major value to the operating facility is as a management tool which provides information on environmental performance in relation to goals and intentions.

ESSENTIAL ELEMENTS OF ENVIRONMENTAL AUDITS

The practice of environmental auditing involves examining critically the operations on a site and, if necessary, identifying areas for improvement to assist the management to meet requirements. The essential steps are the collection of information, the evaluation of that information and the formulation of conclusions including identification of aspects needing improvement. Suggested procedures are described in the Appendix. For environmental auditing to be effective and yield maximum benefit, the following elements are necessary:

1. Full management commitment. It is important that management from the highest levels overtly supports a purposeful and systematic environmental audit programme. Such commitment is demonstrated by, for example, personal interest and concern, the adoption of high standards, the allocation of appropriate manpower and resources, and the active follow-up of recommendations.
2. Audit team objectivity. The principal members of the audit team should be sufficiently detached to ensure objectivity. How this is best arranged will depend upon the size and structure of the company concerned and the nature of the specific audit.
3. Professional competence. Team members should be appropriately qualified and sufficiently senior to provide a technically sound and realistic appraisal, and to command respect. The skills required fall under the headings of general environmental affairs and policy, specific environmental expertise and operational experience and knowledge of environmental auditing.
4. Well-defined and systematic procedures. To ensure comprehensive and efficient coverage of relevant matters, considered procedures such as those outlined in the Appendix should be adopted.
5. Written reports. It is self-evident that the process should be properly documented, and that a clear report should be submitted to the management appropriate to the organization of the company. This report should concentrate on factual, objective observations.
6. Quality assurance. It is desirable to have some mechanism to maintain the quality of the auditing system it-

self, and provide assurance of consistency and reliability to the company.

7. Follow-up. Clearly the full value from auditing can only be obtained if there is active implementation and follow-up of matters identified.

CONCLUSION

The ICC supports and encourages the adoption of environmental auditing programmes by industrial organizations as one element in their environmental management systems. Just as the environmental management systems should reflect the nature of the organization, culture and products of individual businesses, environmental auditing programmes should be individually designed and operated to best meet the specific needs and objectives of the business served.

Experience has demonstrated that the full utility of this management tool can best be achieved if its use is voluntary, and findings are for the exclusive use of company management in carrying out their responsibility to correct all deficiencies promptly.

CHAPTER 14 - APPENDIX: SUGGESTED BASIC STEPS IN AN ENVIRONMENTAL AUDIT

INTRODUCTION

This brief description of suggested basic steps in typical environmental audits summarizes key features, and highlights the well defined and planned structure characteristic of environmental audits wherever they are conducted. There are three essential phases: preparatory pre-audit activities; a site visit normally involving interviews with personnel and inspection of facilities; and post-visit activities.

Environmental audits may have different objectives, be conducted in many different settings by individuals with varied backgrounds and skills, but each audit tends to contain certain common elements. During the audit, a team of individuals completes a field assignment which involves gathering basic facts, analyzing the facts, drawing conclusions concerning the status of the programmes audited with respect to specific criteria, and reporting the conclusions to appropriate management.

These activities are conducted within a formal structure in a sequence that is repeated in each location audited to provide a level of uniformity of coverage and reliability of findings that is maintained from audit to audit. The last page of this chapter gives a typical audit work flow. Although not all audit programmes necessarily contain each step, the design of each programme generally makes provision for each of the activities described.

PRE-AUDIT ACTIVITIES

Preparation for each audit covers a number of activities including selecting the review site and audit team, developing an audit plan which defines the technical, geographic and time scope, and obtaining background information on the plant (for example by means of a questionnaire) and the criteria to be used in evaluating programmes. The intent of these activities is to minimize time spent at the site and to prepare the audit team to operate at maximum productivity throughout the on-site portion of the audit.

With respect to the composition of the audit team, there are both advantages and disadvantages in including a member from the site being audited. Advantages include (i) the insider's knowledge of the specifics of the plant as regards both physical installations and organizational patterns, (ii) associating a local employee with the audit report may make it appear more credible to the plant's workforce. The main disadvantage is that the insider may have difficulty in taking or expressing an objective view, especially if this might be seen as criticism of his superior or immediate colleagues.

Independent consultants may provide assistance, especially to smaller companies, in the event of a lack of internal expertise.

ACTIVITIES AT THE SITE

The audit activities at the site typically include five basic steps:

1. Identifying and understanding management control systems

Internal controls are incorporated in the facility's environmental management system. They include the organizational monitoring and record-keeping procedures, formal planning documents such as plans for prevention and control of accidental release, internal inspection programmes, physical controls such as containment of release material, and a variety of other control system elements. The audit team gains information on all significant control system elements from numerous sources through use of formal questionnaires, observations and interviews.

2. Assessing management control systems

The second step involves evaluating the effectiveness of management control systems in achieving their objectives. In some cases, regulations specify the design of the control system. For example, regulations may list specific elements to be included in plans for responding to accidental releases. More commonly, team members must rely on their own professional judgement to assess adequate control.

3. Gathering audit evidence

In this step the team gathers evidence required to verify that the controls do in practice provide the result intended. Team members follow testing sequences outlined in the audit protocol which have been modified to consider special conditions at the site. Examples of typical tests include review of a sample of effluent monitoring data to confirm compliance with limits, of training records to confirm that appropriate people have been trained, or of purchasing department records to verify that only approved waste disposal contractors have been used. All of the information gathered is recorded for ease of analysis and as a record of conditions at the time of the audit. Where a control element is in some way deficient, the condition is recorded as a "finding".

4. Evaluating audit findings

After the individual controls have been tested and team members have reached conclusions concerning individual elements of the control system, the team meets to integrate and evaluate the findings and to assess the significance of each deficiency or pattern of deficiencies in the overall functioning of the control system. In evaluating the audit findings, the team confirms that there is sufficient evidence to support the findings and summarizes related findings in a way that most clearly

communicates their significance.

5. Reporting audit findings

Findings are normally discussed individually with facility personnel in the course of the audit. At the conclusion of the audit, a formal exit meeting is held with facility management to report fully all findings and their significance in the operation of the control system. The team may provide a written summary to management which serves as an interim report prior to preparation of the final report.

POST-AUDIT ACTIVITIES

At the conclusion of the on-site audit two important activities remain: preparation of the final report and development of a corrective action programme.

1. Final audit report

The final audit report is generally prepared by the team leader and, after review in draft by those in a position to evaluate its accuracy, it is provided to appropriate management.

2. Action plan preparation and implementation

Facility personnel, sometimes assisted by the audit team or outside experts, develop a plan to address all findings. This action plan serves as a mechanism for obtaining management approval and for tracking progress toward its completion. It is imperative that this activity take place as soon as possible so that management can be assured that appropriate corrective action is planned. A primary benefit of the audit is lost, of course, if corrective action is not taken promptly.

Follow-up to ensure that the corrective action plan is carried out and all necessary corrective action is taken is an important step. This may be done by an audit team, by internal environmental experts, or by management.

SUMMARY

Environmental audits, while they may vary in some details, have certain generic characteristics as described above. The approach is characterized by a well-defined and planned structure, careful, methodical investigations and strong emphasis on reporting to all appropriate management. These characteristics from the basis for the reputation that environmental auditing has earned for providing reliable and useful information to management in all settings where it is practiced around the world.

BASIC STEPS OF AN ENVIRONMENTAL AUDIT

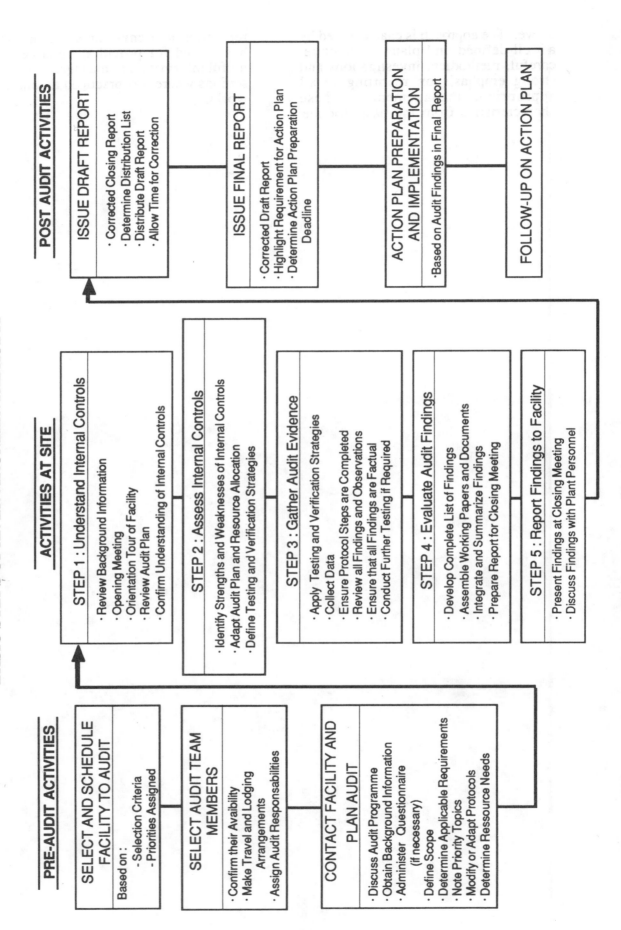

PRE-AUDIT ACTIVITIES

SELECT AND SCHEDULE FACILITY TO AUDIT

Based on :
- Selection Criteria
- Priorities Assigned

SELECT AUDIT TEAM MEMBERS

- Confirm their Availibility
- Make Travel and Lodging Arrangements
- Assign Audit Responsabilities

CONTACT FACILITY AND PLAN AUDIT

- Discuss Audit Programme
- Obtain Background Information
- Administer Questionnaire (if necessary)
- Define Scope
- Determine Applicable Requirements
- Note Priority Topics
- Modify or Adapt Protocols
- Determine Ressource Needs

ACTIVITIES AT SITE

STEP 1 : Understand Internal Controls

- Review Background Information
- Opening Meeting
- Orientation Tour of Facility
- Review Audit Plan
- Confirm Understanding of Internal Controls

STEP 2 : Assess Internal Controls

- Identify Strengths and Weaknesses of Internal Controls
- Adapt Audit Plan and Resource Allocation
- Define Testing and Verification Strategies

STEP 3 : Gather Audit Evidence

- Apply Testing and Verification Strategies
- Collect Data
- Ensure Protocol Steps are Completed
- Review all Findings and Observations
- Ensure that all Findings are Factual
- Conduct Further Testing if Required

STEP 4 : Evaluate Audit Findings

- Develop Complete List of Findings
- Assemble Working Papers and Documents
- Integrate and Summarize Findings
- Prepare Report for Closing Meeting

STEP 5 : Report Findings to Facility

- Present Findings at Closing Meeting
- Discuss Findings with Plant Personnel

POST AUDIT ACTIVITIES

ISSUE DRAFT REPORT

- Corrected Closing Report
- Determine Distribution List
- Distribute Draft Report
- Allow Time for Correction

ISSUE FINAL REPORT

- Corrected Draft Report
- Highlight Requirement for Action Plan
- Determine Action Plan Preparation Deadline

ACTION PLAN PREPARATION AND IMPLEMENTATION

- Based on Audit Findings in Final Report

FOLLOW-UP ON ACTION PLAN

CHAPTER 15 - US ENVIRONMENTAL PROTECTION AGENCY ENVIRONMENTAL AUDITING POLICY STATEMENT

(From the United States Federal Register / Vol. 51. No. 131 / Wednesday July 9,1986)

Summary

It is EPA policy to encourage the use of environmental auditing by regulated entities to help achieve and maintain compliance with environmental laws and regulations, as well as to help identify and correct unregulated environmental hazards. EPA first published this policy as interim guidance on November 8, 1985 (50 FR 46504). Based on comments received regarding the interim guidance, the Agency is issuing today's final policy statement with only minor changes.

This final policy statement specifically:

- encourages regulated entities to develop, implement and upgrade environmental auditing programs;
- discusses when the Agency may or may not request audit reports;
- explains how EPA's inspection and enforcement activities may respond to regulated entities' efforts to assure compliance through auditing;
- endorses environmental auditing at federal facilities;
- encourages state and local environmental auditing initiatives; and
- outlines elements of effective audit programs.

Environmental auditing includes a variety of compliance assessment techniques which go beyond those legally required and are used to identify actual and potential environmental problems. Effective environmental auditing can lead to higher levels of overall compliance and reduced risk to human health and the environment. EPA endorses the practice of environmental auditing and supports its accelerated use by regulated entities to help meet the goals of federal, state and local environmental requirements. However, the existence of an auditing program does not create any defence to, or otherwise limit, the responsibility of any regulated entity to comply with applicable regulatory requirements.

States are encouraged to adopt these or similar and equally effective policies in order to advance the use of environmental auditing on a consistent nationwide basis.

Dates: This final policy statement is effective July 9. 1986.
For further information contact:
Office of Policy Planning and Evaluation (202) 382—2726
or
Office of Enforcement and Compliance Monitoring (202) 382—7550

ENVIRONMENTAL AUDITING POLICY STATEMENT

I. PREAMBULE

On November 8, 1985 EPA published an Environmental Auditing Policy Statement, effective as interim guidance, and sollicited written comments until January 7, 1986.

Thirteen commenters submitted written comments. Eight were from private industry. Two commenters represented industry trade associations. One federal agency, one consulting firm and one law firm also submitted comments.

Twelve commenters addressed EPA requests for audit reports. Three comments per subject were received regarding inspections enforcement response and elements of effective environmental auditing. One commenter addressed audit provisions as remedies in enforcement actions, one addressed environmental auditing at federal facilities, and one addressed the relationship of the policy statement to state or local regulatory agencies. Comments generally supported both the concept of a policy statement and the interim guidance, but raised specific concerns with respect to particular language and policy issued in sections of the guidance.

General Comments

Three commenters found the interim guidance to be constructive, balanced and effective at encouraging more and better environmental auditing.

Another commenter, while considering the policy on the whole to be constructive felt that new and identifiable auditing "incentives" should be offered by EPA. Based on earlier comments received from industry, EPS believes most companies would not support or participate in an "incentives–based" environmental auditing program with EPA. Moreover, general promises to forgo inspections or reduce enforcement responses in exchange for companies' adoption of environmental auditing programs — the "incentives" most frequently mentioned in this context — are fraught with legal and policy obstacles.

Several commenters expressed concern that states or localities might use the interim guidance to require auditing. The Agency disagrees that the policy statement opens the way for states and localities to require auditing. No EPA policy can grant states or localities any more (or less) authority than they already possess. EPA believes that the interim guidance effectively encourages voluntary auditing. In fact, Section II.B of the policy states: "because audit quality depends to a large degree on genuine management commitment to the program and its objectives, auditing should remain a voluntary program".

Another commenter suggested that EPA should not expect an audit to identify all potential problem areas or conclude that a problem identified in an audit reflects normal operations and procedures. EPA agrees than an audit report should clearly reflect these realities and should be written to point out the audit's limitations. However, since EPA will not routinely request audit reports, the Agency does not believe these concerns raise issues which need to be addressed in the policy statement.

A second concern expressed by the same commenter was that EPA should acknowledge that environmental audits are only part of a successful environmental management program and thus should not be expected to cover every environmental issue or solve all problems. EPA agrees and accordingly has amended the statement of purpose which appears at the end of the preamble.

Yet another commenter thought EPA should focus on environmental performance results (compliance or non-compliance), not on the processes or vehicles used to achieve those results. In general, EPA agreed with this statement and will continue to focus on environmental results. However, EPA also believes that such results can be improved through Agency efforts to identify and encourage effective environmental management practices and will continue to encourage such practices in non-regulatory ways.

A final general comment recommended that EPA should sponsor seminars for small businesses on how to start auditing programs. EPA agrees that such seminars would be useful.

However, since audit seminars already are available from several private sector organizations, EPA does not believe it should intervene in that market with the possible exception of seminars for government agencies, especially federal agencies, for which EPA has a broad mandate under Executive Order 12068 to provide technical assistance for environmental compliance.

Requests for reports

EPA received 12 comments regarding Agency requests for environmental audit reports, far more than on any other topic in the policy statement. One commenter felt that EPA struck an appropriate balance between respecting the need for self-evaluation with some measure of privacy, and allowing the Agency enough flexibility of inquiry to accomplish future statutory missions. However, most commenters expressed concern that the interim guidance did not go far enough to assuage corporate fears that EPA will use audit reports for environmental compliance "witch hunts". Several commenters suggested additional specific assurances regarding the circumstances under which EPA will request such reports.

One commenter recommended that EPA request audit reports only "when the Agency can show the information it needs to perform its statutory mission cannot be obtained from the monitoring, compliance or other data that is otherwise reportable and/or accessible to EPA or where the Government deems an audit report material to a criminal investigation". EPA accepts this recommendation in part. The Agency believes it would not be in the best interest of human health and the environment to commit to making a "showing" of a compelling information need before ever requesting an audit report. While EPA may normally be willing to do so, the Agency cannot rule out in advance all circumstances in which such a showing may not be

possible. However, it would be hepful to further clarify that a request for an audit report or a portion of a report normally will be made when needed information is not available by alternative means. Therefore, EPA has revised section III.A paragraph two and added the phrase: "and usually made where the information needed cannot be obtained from monitoring, reporting or other data otherwise available to the Agency".

Another commenter suggested that (except in the case of criminal investigations) EPA should limit requests for audit documents to specific questions. By including the phrase "or relevant portions of a report" in Section III.A, EPA meant to emphasize it would not request are entire audit document when only a relevant portion would suffice. Likewise, EPA fully intends not to request even a portion of a report if needed information or data can be otherwise obtained. To further clarify this point EPA has added the phrase "most likely focused on particular information need rather than the entire report" to the second sentence of paragraph two, Section III A.

Incorporating the two comments above, the first two sentences in paragraph two of final Section III.A now read: "EPA's authority to request an audit report, or relevant portions thereof, will be exercised on a case-by-case basis where the Agency determines it is needed to accomplish a statutory mission or the Government deems it to be material to a criminal investigation. EPA expects such requests to be limited, most likely focused on particular information needs rather than the entire report, and usually made where the information needed cannot be obtained from monitoring, reporting or other data otherwise available to the Agency".

Other commenters recommended that EPA not request audit reports under any circumstances, that requests be "restricted to only those legally required" that requests be limited to criminal investigations, or that requests be made only when EPA has reason to believe "that the audit programs or reports are being used to conceal evidence of environmental non-compliance or otherwise being used in bad faith." EPA appreciates concerns underlying all of these comments and has considered each carefully. However the Agency believes that these recommendations do not strike the appropriate balance between retaining the flexibility to accomplish EPA's statutory missions in future, unforeseen circumstances and acknowledging regulated entities' need to self-evaluate environmental performance with some measure of privacy. Indeed, based on prime informal comments, the small number of comments received and the even smaller number of adverse comments, EPA believes the final policy statement should remain largely unchanged from the interim version.

Elements of effective environmental auditing

Three commenters expressed concerns regarding the seven general elements EPA outlined in the Appendix to the interim guidance.

One commenter noted that were EPA to further expand or more fully detail such elements, programs not specifically fulfilling each element would then be judged inadequate. EPA agrees that presenting highly specific and prescriptive auditing elements could be counter-productive by not taking into account numerous factors which vary extensively from one organization to another, but which may still result in effective auditing programs.

Accordingly, EPA does not plan to expand or more fully detail these auditing elements.

Another commenter asserted that states and localities should be cautioned not to consider EPA's auditing elements as mandatory steps. The Agency is fully aware of this concern and in the interim guidance noted its strong opinion that "regulatory agencies should not attempt to prescribe the precise form and structure of regulated entities environmental management or auditing programs". While EPA cannot require state or local regulators to adopt this or similar policies, the Agency does strongly encourage them to do so, both in the interim and final policies.

A final commenter thought the Appendix too specifically prescribed what should and what should not be included in an auditing program. Other commenters, on the other hand, viewed the elements described as very general in nature. EPA agrees with these other commenters. The elements are in no way binding. Moreover, EPA believes that most mature, effective environmental auditing programs do incorporate each of these general elements in some form, and considers them useful yardsticks for those considering adopting or upgrading audit programs. For these reasons EPA has not revised the Appendix in today's final policy statement.

Other comments

Other significant comments addressed EPA inspection priorities for, and enforcement responses to, organizations with environmental auditing programs.

One commenter, stressing that audit programs are internal management tools, took exception to the phrase in the second paragraph of Section III.B.1 of the interim guidance which states that environmental audits can "complement" regulatory oversight. By using the word "complement" in this context, EPA does not intend to imply that audit reports must be obtained by the agency in order to supplement regulatory inspections. "Complement" is used in a broad sense of being in addition to inspections and providing something (i.e. self-assessment) which otherwise would be lacking. To clarify this point EPA has added the phrase "by providing self-assessment to assure compliance" after "environmental audits may complement inspections" in this paragraph.

The same commenter also expressed concern that, as EPA sets inspection priorities, a company having an audit program could appear to be a "poor performer" due to complete and accurate reporting when measured against a company which reports something less than required by law. EPA agrees that it is important to communicate this fact to Agency and state personnel, and will do so. However, the Agency does not believe a change in the policy statement is necessary.

A further comment suggested EPA should commit to take auditing programs into account when assessing all enforcement actions. However, in order to maintain enforcement flexibility under varied circumstances, the Agency cannot promise reduced enforcement responses to violations at all audited facilities when other factors may be overriding. Therefore the policy statement continues to state that EPA may exercise its discretion to consider auditing programs as evidence of honest and genuine efforts to assure compliance, which would then be taken into account in fashioning enforcement responses to violations.

A final commenter suggested the phrase "expeditiously correct environmental problems" not be used in the enforcement context since it implied EPA would use an entity's record of correcting nonregulated matters when evaluating regulatory violations. EPA did not intend for such an inference to be made. EPA intended the term "environmental problems" to refer to the underlying circumstances which eventually lead up to the violations. To clarify this point, EPA is revising the first two sentences of the paragraph to which this comment refers by changing "environmental problems" to "violations and underlying environmental problems" in the first sentence and to "underlying environmental problems" in the second sentence.

In a separate development EPA is preparing an update of its January 1984 Federal Facilities Compliance Strategy, which is referenced in section III.C. of the auditing policy. The Strategy should be completed and available on request from EPA's Office of Federal Activities later this year.

EPA thanks all commenters for responding to the November 8, 1985 publication. Today's notice is being issued to inform regulated entities and the public of EPA's final policy toward environmental auditing. This policy was developed to help (a) encourage regulated entities to institutionalize effective audit practices as one means of improving compliance and sound

environmental management, and (b) guide internal EPA actions directly related to regulated entities' environmental auditing programs.

EPA will evaluate implementation of this final policy to ensure it meets the above goals and continues to encourage better environmental management while strengthening the Agency's own efforts to monitor and enforce compliance with environmental requirements.

II. GENERAL EPA POLICY ON ENVIRONMENTAL AUDITING

A. Introduction

Environmental auditing is a systematic documented periodic and objective review by regulated entities[1] of facility operations and practices related to meeting environmental requiremements. Audits can be designed to accomplish any or all of the following: verify compliance with environmental requirements; evaluate the effectiveness of environmental management systems already in place; or assess risk from regulated and unregulated materials and practices.

Auditing serves as a quality assurance check to help improve the effectiveness of basic environmental management by verifying that management practices are in place, functioning and adequate. Environmental audits evaluate, and are not a substitute for, direct compliance activites such as obtaining permits, installing controls, monitoring compliance, reporting violations, and keeping records. Environmental auditing may verify but does not include activities required by law, regulation or permit (e.g. continuous emissions monitoring, composite correction plans at wastewater treatment plants, etc.).Audits do not in any way replace regulatory

---

[1] "Regulated entities include private firms and public agencies with facilities subject to environmental regulations. Public agencies can include federal, state or local agencies as well as special purpose organizations such as regional sewage commissions.

agency inspections. However environmental audits can improve compliance by complementing conventional federal, state and local oversight.

The appendix to this policy statement outlines some basic elements of environmental auditing (e.g. auditor independence and top management support) for use by those considering implementation of effective auditing programs to help achieve and maintain compliance. Additional information on environmental auditing practices can be found in various published materials[2].

Environmental auditing has developed for sound business reasons, particularly as a means business reasons, particularly as a means of helping regulated entities manage pollution control affirmatively over time instead of reacting to crises. Auditing can result in improved facility environmental performance, help communicate effective solutions to common environmental problems, focus facility managers' attention on current and upcoming regulatory requirements and generate protocols and checklists which help facilities better manage themselves. Auditing also can result in better-integrated management of environmental hazards, since auditors frequently identify environmental liabilities which go beyond regulatory compliance. Companies, public entities and federal facilities have employed a variety of environmental auditing practices in recent years. Several hundred major firms in diverse industries now have environmental auditing programs, although they often are known by other names such as assessment survey, surveillance, review or appraisal.

While auditing has demonstrated its usefulness to those with audit programs many others still do not audit.

Clarification of EPA's position re-

---

2 See. e.g. "Current Practices in Environmental Auditing." EPA Report N°EPA-230-09-83-006. February 1984: "Annotated Bibliography on Environmental Auditing." Fifth Edition, September 1985, both available from: Regulatory Reform Staff, PM-223. EPA. 401 M Street SW Washington DC 20460

garding auditing may help encourage regulated entities to establish audit programs or upgrade systems already in place.

B. EPA Encourages the Use of Environmental Auditing

EPA encourages regulated entities to adopt sound environmental management practices to improve environmental performance. In particular, EPA encourages regulated entities subject to environmental regulations to institute environmental auditing programs to help ensure the adequacy of internal systems to achieve, maintain and monitor compliance. Implementation of environmental problems, as well as improvements to management practices. Audits can be conducted effectively by independent internal or third party auditors. Larger organizations generally have greater resources to devote to an internal audit team, while smaller entities might be more likely to use outside auditors.

Regulated entities are responsible for taking all necessary steps to ensure compliance with environmental requirements, whether or not they adopt audit programs. Although environmental laws do not require a regulated facility to have an auditing program, ultimate responsibility for the environmental performance of the facility lies with top management, which therefore has a strong incentive to use reasonable means, such as environmental auditing, to secure reliable information of facility compliance status.

EPA does not intend to dictate or interfere with the environmental management practices of private or public organizations. Nor does EPA intend to mandate auditing (though in certain instances EPA may seek to include provisions for environmental auditing as part of settlement agreements, as noted below). Because environmental auditing systems have been widely adopted on an voluntary basis in the past, and because audit quality depends to a large degree upon genuine management commitment to the program and its objectives, auditing should remain a voluntary activity.

III. EPA POLICY ON SPECIFIC ENVIRONMENTAL AUDITING ISSUES

A. Agency Requests for Audit Reports

EPA has broad statutory authority to request relevant information on the environmental compliance status of regulated entities However, EPA believes routine Agency requests for audit reports[3] could inhibit auditing in the long run, decreasing both the quantity and quality of audits conducted. Therefore, as a matter of policy, EPA will not routinely request environmental audit reports.

EPA's authority to request an audit report, or relevant portions thereof, will be exercised on a case-by-case basis where the Agency determines it is needed to accomplish a statutory mission, or where the Government deems it to be material to a criminal investigation. EPA expects such requests to be limited, most likely focused on particular information needs rather than the entire report, and usually made where the information needed cannot be obtained from monitoring, reporting or other data otherwise available to the Agency. Examples would likely include situations where audits are conducted under consent decrees or other settlement agreements; a company has placed its management practices at issue by raising them as a defence; or state of mind or intent are a relevant element of inquiry such as during a criminal investigation. This list is illustrative rather than exhaustive since there doubtless will be other situations, not subject to prediction, in which audit reports rather than information may be required.

EPA acknowledges regulated entities' need to self-evaluate environmental performance with some measure of privacy and encourages such activity. However, audit reports may not shield monitoring, compliance, or other information that would otherwise be reportable and/or accessible to EPA, even if there is no explicit "requirement" to generate that data[4]. Thus, this policy does not alter regulated entities' existing or future obligations to monitor, record or report information required under environmental statutes, regulations or permits, or to allow EPA access to that information. Nor does this policy alter EPA's authority to request and receive any relevant information — including that contained in audit reports — under various environmental statutes (e.g. Clean Water Act Section 308. Clean Air Act sections 114 and 208) or in other administrative or judicial proceedings.

Regulated entities also should be aware that certain audit findings may by law have to be reported to government agencies. However, in addition to any such requirements, EPA encourages regulated entities to notify appropriate State or Federal Officials of findings which suggest significant environmental or public health risks, even when not specifically required to do so.

B. EPA Response to Environmental Auditing

1. General Policy

EPA will not promise to forgo inspections, reduce enforcement responses, or offer other such incentives in exchange for implementation of environmental auditing or other sound environmental management practices. Indeed, a credible enforcement program provides a strong incentive for regulated entities to audit.

Regulatory agencies have an obligation to assess source compliance status independently and cannot eliminate in-

---

[3] An "environmental audit report" is a written report which candidly and thoroughly presents findings from a review, conducted as part of an environmental audit as described in section II.A., of facility environmental performance and practices. An audit report is not a substitute for compliance monitoring reports or other reports or records which may be requirred by EPA or other regulatory agencies.

[4] See, for example, "Duties to Report or Disclose Information on the Environmental Aspects of Business Activities." Environmental Law Institute report to EPA, final report, September 1985.

spections for particular firms or classes of firms. Although environmental audits may complement inspections by providing self-assessment to assure compliance, they are in no way a substitute for regulatory oversight. Moreover, certain statutes (e.g. RCRA) and Agency policies establish minimum facility inspection frequencies to which EPA will adhere.

However EPA will continue to address environmental problems on a priority basis and will consequently inspect facilities with poor environmental records and practices more frequently. Since effective environmental auditing helps management identify and promptly correct actual or potential problems, audited facilities, environmental performance should improve. Thus, while EPA inspections of self-audited facilities will continue, to the extent that compliance performance is considered in setting inspection priorities, facilities with a good compliance history may be subject fo fewer inspections.

In fashioning enforcement responses to violations, EPA policy is to take into account, on a case-by-case basis, the honest and genuine efforts of regulated entities to avoid and promptly correct violations and underlying environmental problems. When regulated entities take reasonable precautions to avoid non-compliance, expeditiously correct underlying environmental problems discovered through audits or other means, and implement measures to prevent their recurrence. EPA may exercise its discretion to consider such actions as honest and genuine efforts to assure compliance. Such consideration applies particularly when a regulated entity promptly reports violations or compliance data which otherwise were not required to be recorded or reported to EPA.

2. Audit Provisions as Remedies in Enforcement Actions

EPA may propose environmental auditing provisions in consent decrees and in other settlement negotiations where auditing could provide a remedy for identified problems and reduce the likelihood of similar problems recurring in the future[5]. Environmental auditing provisions are most likely to be proposed in settlement negotiations where:

- a pattern of violations can be attributed at least in part, to the absence or poor functioning of an environmental management system; or
- the type or nature of violations indicates a likelihood that similar non-compliance problems may exist or occur elsewhere in the facility or at other facilities operated by the regulated entity.

Through this consent decree approach and other means, EPA may consider how to encourage effective auditing by publicly owned sewage treatment works (POTWs). POTWs often have compliance problems related to operation and maintenance procedures which can be addressed effectively through the use of environmental auditing. Under its National Municipal Policy EPA already is requiring many POTWs to develop composite correction plans to identify and correct compliance problems.

C. Environmental Auditing of Federal Facilities

EPA encourages all federal agencies subject to environmental laws and regulations to institute environmental auditing systems to help ensure the adequacy of internal systems to achieve, maintain and monitor compliance. Environmental auditing at federal facilities can be an effective supplement to EPA and state inspections. Such federal facility environmental audit programs should be structured to promptly identify environmental problems and expeditiously develop schedules for remedial action.

To the extent feasible, EPA will provide technical assistance to help federal agencies design and initiate audit pro-

---

[5] EPA is developing guidance for use by Agency negotiators in structuring appropriate environmental audit provisions for consent decrees and other settlement negotiations.

grams. Where appropriate, EPA will enter into agreements with other agencies to clarify the respective roles, responsibilities and commitments of each agency in conducting and responding to federal facility environmental audits.

With respect to inspections of self-audited facilities (see section III.B.1 above) and requests for audit reports (see section III.A above), EPA generally will respond to environmental audits by federal facilities in the same manner as it does for other regulated entities, in keeping with the spirit and intent of Executive Order 12088 and the EPA Federal Facilities Compliance Strategy (January 1984, update forthcoming in late 1986). Federal agencies should, however, be aware that the Freedom of Information Act will govern any disclosure of audits reports or audit-generated information requested from federal agencies by the public.

When federal agencies discover significant violations through an environmental audit, EPA encourages them to submit the related audit findings and remedial action plans expeditiously to the applicable EPA regional office (and responsible state agencies, where appropriate) even when not specifically required to do so. EPA will review the auditing findings and action plans and either provide written approval or negotiate a Federal Facilities Compliance Agreement. EPA will utilize the escalation procedures provided in Executive Order 12088 and the EPA Federal Facilities Compliance Strategy only when agreement between agencies cannot be reached. In any event, federal agencies are expected to report pollution abatement projects involving costs (necessary to correct problems discovered through the audit) to EPA in accordance with OMB Circular A–106. Upon request and in appropriate circumstances, EPA will assist affected federal agencies through coordination of any public release of audit findings with approved action plans once agreement has been reached.

IV. RELATIONSHIP TO STATE OR LOCAL REGULATORY AGENCIES

State and local regulatory agencies have independent jurisdiction over regulated entities. EPA encourages them to adopt these or similar policies, in order to advance the use of effective environmental auditing in a consistent manner.

EPA recognizes that some states have already undertaken environmental auditing initiatives which differ somewhat from this policy. Other states also may want to develop auditing policies which accommodate their particular needs or circumstances. Nothing in this policy statement is intended to preempt or preclude states from developing other approaches to environmental auditing. EPA encourages state and local authorities to consider the basic principles which guided the Agency in developing this policy:

- Regulated entities must continue to report or record compliance information required under existing statutes or regulations, regardless of whether such information is generated by an environmental audit or contained in an audit report. Required information cannot be withheld merely because it is generated by an audit rather than by some other means.
- Regulatory agencies cannot make promises to forgo or limit enforcement action against a particular facility or class of facilities in exchange for the use of environmental auditing systems. However, such agencies may use their discretion to adjust enforcement actions on a case-by-case basis in response to honest end genuine efforts by regulated entities to assure environmental compliance.
- When setting inspection priorities regulatory agencies should focus to the extent possible on compliance performance and environmental results.
- Regulatory agencies must continue to meet minimum program requirements (e.g. minimum inspection requirements, etc.).

– Regulatory agencies should not attempt to prescribe the precise form and structure of regulated entities' environmental management or auditing programs.

An effective state/federal partnership is needed to accomplish the mutual goal of achieving and maintaining high levels of compliance with environmental laws and regulations. The greater the consistency between state or local policies and this federal response to environmental auditing, the greater the degree to which sound auditing practices might be adopted and compliance levels improve.

Dated: June 28, 1986
Lee M. Thomas
Administrator:

CHAPTER 15 - APPENDIX
ELEMENTS OF EFFECTIVE ENVIRONMENTAL AUDITING PROGRAMS

INTRODUCTION

Environmental auditing is a systematic, documented, periodic and objective review by a regulated entity of facility operations and practices related to meeting environmental requirements.

Private sector environmental audits of facilities have been conducted for several years and have taken a variety of forms, in part to accommodate unique organizational structures and circumstances. Nevertheless, effective environmental audits appear to have certain discernible elements in common with other kinds of audits. Standards for internal audits have been documented extensively. The elements outlined below draw heavily on two of these documents: "Compendium of Audit Standards (©1983, Walter Willborn, American Society for Quality Control) and "Standards for the Professional Practice of Internal Auditing" (©1981, The Institute of Internal Auditors, Inc.). They also reflect Agency analyses conducted over the last several years.

Performance-oriented auditing elements are outlined here to help accomplish several objectives. A general description of features of effective, mature audit programs can help those starting audit programs, especially federal agencies and smaller businesses. These elements also indicate the attributes of auditing EPA generally considers important to ensure program effectiveness. Regulatory agencies may use these elements in negotiating environmental auditing provisions for consent decrees. Finally, these elements

can help guide states and localities considering auditing initiatives.

An effective environmental auditing system will likely include the following general elements:

I.   *Explicit top management support for environmental auditing and commitment to follow-up on audit findings.* Management support may be demonstrated by a written policy articulating upper management support for the auditing program, and for compliance with all pertinent requirements, including corporate policies and permit requirements as well as federal, state and local statutes and regulations.

Management support for the auditing program also should be demonstrated by an explicit written commitment to follow up on audit findings to correct identified problems and prevent their recurrence.

II.   *An environmental auditing function independent of audited activities.* The status or organizational locus of environmental auditors should be sufficient to ensure objective and unobstructed inquiry, observation and testing. Auditor objectivity should not be impaired by personal relationships financial or other conflicts of interest, interference with free inquiry or judgement or fear of potential retribution.

III.   *Adequate team staffing and auditor training.* Environmental auditors should possess or have ready access to the knowledge, skills and disciplines needed to accomplish audit objectives.

Each individual auditor should comply with the company's professional standards of conduct. Auditors, whether full-time or part-time, should maintain their technical and analytical competence through continuing education and training.

IV. *Explicit audit program objectives, scope, resources and frequency.* At a minimum, audit objectives should include assessing compliance with applicable environmental laws and evaluating the adequacy of internal compliance policies, procedures and personnel training programs to ensure continued compliance.

Audits should be based on a process which provides auditors: all corporate policies, permits, and federal, state and local regulations pertinent to the facility; and checklists or protocols addressing specific features that should be evaluated by auditors.

Explicit written audit procedures generally should be used for planning audits, establishing audit scope, examining and evaluating audit findings, communicating audit results, and following up.

V. *A process which collects, analyzes, interprets and documents information sufficient to achieve audit objectives.* Information should be collected before and during an onsite visit regarding environmental compliance [1], environmental management effectiveness [2], and other matters [3] related to audit objectives and scope. This information should be sufficient, reliable, relevant and useful to provide a sound basis for audit findings and recommendations.

a. *Sufficient* information is factual, adequate and convincing so that a prudent, informed person would be likely to reach the same conclusions as the auditor.

b. *Reliable* information is the best attainable through use of appropriate audit techniques.

c. *Relevant* information supports audit findings and recommendations and is consistent with the objectives for the audit.

d. *Useful* information helps the organization meet its goals.

The audit process should include a periodic review of the reliability and integrity of this information and the means used to identify, measure, classify and report it. Audit procedures, including the testing and sampling techniques employed, should be selected in advance, to the extent practical, and expanded or altered if circumstances warrant. The process of collecting, analyzing, interpreting and documenting information should provide reasonable assurance that audit objectivity is maintained and audit goals are met.

VI. *A process which includes specific procedures to promptly prepare candid, clear and appropriate written reports on audit findings, corrective actions, and schedules for implementation.* Procedures should be in place to ensure that such information is communicated to managers, including facility and corporate management, who can evaluate the information and ensure correction of identified problems.

Procedures also should be in place for determining what internal findings are reportable to state or federal agencies.

VII. *A process which includes quality assurance procedures to assure the accuracy and thoroughness of environmental audits.* Quality assurance may be accomplished through supervision, independent internal reviews, external reviews or a combination of these approaches.

Footnotes to Appendix

(1) A comprehensive assessment of compliance with federal environmental regulations requires an analysis of facility performance against numerous environmental statutes and implementing regulations. These statutes include:

Resource Conservation and Recovery Act
Federal Water Pollution Control Act
Clean Air Act
Hazardous Materials Transportation Act
Toxic Substances Control Act
Comprehensive Environmental Response, Compensation and Liability Act
Safe Drinking Water Act
Federal Insecticide, Fungicide an Rodenticide Act
Marine Protection, Research and Sanctuaries Act
Uranium Mill Tailings Radiation Control Act

In addition, state and local government are likely to have their own environmental laws. Many states have been delegated authority to administer federal programs. Many local governments' building, fire, safety and health codes also have environmental requirements relevant to an audit evaluation.

(2) An environmental audit could go well beyond the type of compliance assessment normally conducted during regulatory inspections, for example, by evaluating policies and practices, regardless of whether they are part of the environmental systems or the operating and maintenance procedures. Specifically, audits can evaluate the extent to which systems or procedures:

1. Develop organizational environmental policies which: a. implement regulatory requirements; b. provide management guidance for environmental hazards not specifically addressed in regulations;
2. Train and motivate facility personnel to work in an environmentally-acceptable manner and to understand and comply with government regulations and the entity's environmental policy;
3. Communicate relevant environmental developments expeditiously to facility and other personnel;
4. Communicate effectively with government and the public regarding serious environmental incidents;
5. Require third parties working for, with or on behalf of the organization to follow its environmental procedures;
6 Make proficient personnel available at all times to carry out environmental (especially emergency) procedures;
7. Incorporate environmental protection into written operating procedures;
8. Apply best management practices and operating procedures, including "good housekeeping" techniques;
9. Institute preventive and corrective maintenance systems to minimize actual and potential environmental harm;
10. Utilize best available process and control technologies;
11. Use most-effective sampling and monitoring techniques, test methods, recordkeeping systems or reporting protocols (beyond minimum legal requirements);
12. Evaluate causes behind any serious environmental incidents and establish procedures to avoid recurrence;
13. Exploit source reduction, recycle and reuse potential wherever practical; and
14. Substitute materials or processes to allow use of the least-hazardous substances feasible.

(3) Auditors could also assess environmental risks and uncertainties.

CHAPTER 16 - MEASURES TO PROMOTE COMPLIANCE: PROMOTION OF ENVIRONMENTAL AUDITS

Extracts from:
"Enforcement and Compliance Policy –
Canadian Environmental Protection Act"
Published by Environment Canada
May 1988

Environmental audits are internal evaluations by companies and Government agencies to verify their compliance with legal requirements as well as their own internal policies and standards. They are conducted by companies, government agencies and others on a voluntatry basis, and are carried out by outside consultants or employees of the company or facility from outside the work unit being audited. Audits can identify compliance problems, weaknesses in management systems, or areas of risk. The findings are documented in a written report.

Environment Canada recognizes the power and effectiveness of environmental audits as a management tool for companies and government agencies, and intends to promote their use by industry and others.

To encourage the practice of environmental auditing, inspections and investigations under the *Canadian Environmental Control Act* will be conducted in a manner which will not inhibit the practice or quality of auditing. Inspectors will not request environmental audit reports during routine inspections to verify compliance with the Act.

Access to environmental audit reports may be required when inspectors or investigation specialists have reasonnable grounds to believe that:

– an offence has been committed;
– the audit's findings will be relevant to the particular violation, necessary to its investigation and required as evidence;
– the information being sought through the audit cannot be obtained from other sources through the exercise of the inspector's or investigation specialist's powers.

In particular reference to the latter criterion, environmental audit reports must not be used to shelter monitoring, compliance or other information that would otherwise be accessible to inspectors under the *Canadian Environmental Protection Act.*

Any demand for access to environmental audit reports during investigations will be made under the authority of a search warrant. The only exception to the use of a search warrant is exigent circumstances, that is, when the delay necessary to obtain a warrant would likely result in danger to the environment or human life, or the loss or destruction of evidence.

ANNEX 1
A FEW USEFUL REFERENCES

The Environmental Audit Handbook Series (5 volumes) Executive Enterprises Publications Co., NewYork, 1988.

Environmental Audits, 5th edition, Lawrence B.Cahill (ed.). Government Institutes, Rockville, USA, 1987.

Environmental Auditing - Fundamentals and Techniques, 2nd Edition, by J. Ladd Greene and others. Centre for Environmental Asssurance, Arthur D. Little Inc., Cambridge, Mass., USA, 1988.

Benefits to Industry of Environmental Auditing. Centre for Environmental Insurance, Arthur D. Little Inc., Cambridge, Mass., USA, 1983.

Compendium of Audits Standards, by Walter Willborn. American Society for Quality Control, Milwaukee, Wisconsin, USA, 1983.

Annotated Bibliography on Environmental Auditing. Office of Policy Planning and Evaluation, US EPA, Washington D.C., USA, 1988.

UNEP/IEO Industry and Environment Review,, Vol. 11, N° 4 (1988).

ANNEX II
AGENDA

Tuesday, 10 January 1989

9.30   Opening Remarks by Ms J. Aloisi de Larderel

9.45   Presentation by Mr. Jonathan Plaut, Allied Signal

10.45 Presentation by Mr. Per A. Syrrist, Norsk Hydro A.S.

11.45 Presentation by Ms. Linda A. Woolley, ITT Corporation

12.45 Working lunch in the office

14.00 Presentation by Dr. I.J. Graham-Bryce, Shell International Petroleum Maatschappij B.V.

15.00 Presentation by Mr. Michael L. Kinworthy, Unocal Corporation

16.00 Presentation by Mr. Eric B. Cowell, BP International Ltd.

17.00 Presentation by Mr. Richard Almgren, Federation of Swedish Industry

18.00 Cocktail

Wednesday, 11 January 1989

9.00   Presentation by Mr. Hennie Veldhuizen, Noranda Inc.

10.00 Presentation by Mr. Jacques Salamitou, Rhône Poulenc

11.00 Coffee Break

11.15 Presentation by Dr. G. Eigenmann, Ciba Geigy

12.15 Presentation by Dr. Egon Keller, Ecosystem

13.15 Working lunch in the office

14.30 General Discussion

17.30 End of the Meeting

ANNEX III
PARTICIPANTS LIST

Mr. Richard Almgren

Associate Head
Environment and Natural Ressources Department
Federation of Swedish Industries
Box 5501
114 85 Stockholm
Sweden

tel: 46.8.783 80 00
telex: 19990
fax: 46.8.662 35 95

Mr. Nigel Blackburn

Secretary, Commission on Environment
International Chamber of Commerce
38 cours Albert 1er
75008 Paris
France

tel: 33.1.49 53 28 28
telex: 650770
fax: 33.1. 42 25 86 63

Mr. Eric B. Cowell

Manager, Biological Service
BP International Ltd.
Group Environmental Services
Britannic House, Moor Lane
London EC2Y 9BU
United Kingdom

tel: 44.1.920 64 46
telex: 888811
fax: 44.1.920 82 63

Dr. G. Eigenmann

Audit and Staff Environment
Ciba-Geigy AG
4002 Basel
Switzerland

tel: 41.61.696 71 27
fax: 41.61.696 45 43

Mr. I.J. Graham-Bryce

Health and Safety and Environment Division
Shell Internationale Petroleum
Maatschappij B.V.
Postbus 162
2501 The Hague
The Netherlands

tel: 31.70.77 91 11
telex: 36000
fax: 31.70.77 48 48

Dr. Egon Keller

Managing Director
Ecosystem
Goethestr. 43
4005 Meerbush 2
Federal Republic of Germany

tel: 49.2159.41 35
fax: 49.2159.54 77

Mr. Michael L. Kinworthy

Manager of Environmental Programs
Corporate Environmental Department
Unocal Corporation
1201 West 5th Street
P.O. Box 7600
Los Angeles, CA 90051
USA

tel: 1.213.977 68 10
fax: 1.213.977 63 64

Mr. Jonathan Plaut

Environmental Compliance
Allied Signal Inc.
Health, Safety and Environmental Sciences
P.O. Box 10113R
Morriston, NJ 07960 - 1013
USA

tel: 1.201.455 65 70
fax: 1.201.455 48 35

Mr. Jacques Salamitou

Délégué à l'environnement
Rhône-Poulenc
25 quai Paul Doumer
92408 Courbevoie Cedex
France

tel: 33.1.47 68 11 28
telex: 610500
fax: 33.1.47 68 29 15

Mr. Per Arne Syrrist

Chief Engineer
Norsk Hydro A.S.
P.B. 2594 Solli
0203 Oslo 2
Norway

tel: 47.2.43 21 00
telex: 78350
fax: 47.2.43 27 25

Mr. Hennie Veldhuizen

Director, Environment
Noranda Inc.
4 King St. West
Toronto, Ontario
Canada M5H 3X2

tel: 1.416.982 73 07
fax: 1.416.982 70 21

Ms. Linda A. Woolley

Director, Public Affairs
ITT Corporation
1600 M Street, N.W.
Washington, D.C. 20036
USA

tel: 1.202.296 60 00
fax: 1.202.296 30 55

UNEP SECRETARIAT:

Mrs. J. Aloisi de Larderel

Director
UNEP Industry and Environment Office
Tour Mirabeau
39/43 quai André Citroën
75739 Paris Cedex 15
France

tel: 33.1.40 58 88 58
telex: 204997
fax: 33.1.40 58 88 74

Mr. de la Perrière

Consultant

Lay-out by TWIGA s.a.r.l.

Printer : GAUTHIER-VILLARS
1, boul. Ney. 75018 Paris
Printed in France